Operational Research for Health Policy: Making Better Decisions

Sally Brailsford
Paul Harper
(eds)

Operational Research for Health Policy: Making Better Decisions

Proceedings of the
31st Annual Conference
of the European Working Group
on Operational Research Applied
to Health Services

PETER LANG
Oxford • Bern • Berlin • Bruxelles • Frankfurt am Main • New York • Wien

Bibliographic information published by Die Deutsche Bibliothek
Die Deutsche Bibliothek lists this publication in the Deutsche Nationalbibliografie; detailed bibliographic data is available on the Internet at ‹http://dnb.ddb.de›.

British Library and Library of Congress Cataloguing-in-Publication Data:
A catalogue record for this book is available from *The British Library*, Great Britain, and from *The Library of Congress*, USA

ISBN 978-3-03911-093-3

Hochfeldstrasse 32, Postfach 746, CH-3000 Bern 9, Switzerland
info@peterlang.com, www.peterlang.com, www.peterlang.net

PREFACE

The EURO Working Group on Operational Research Applied to Health Services (ORAHS) was formed in 1975 as part of the programme for developing special interest groups within the European branch, EURO, of the International Federation of OR Societies (IFORS). At present the group has over 170 members from around 25 countries, mainly in Europe, although ORAHS also has members in Canada, Brazil, Australia, Japan and the USA. The group meets every year, each summer, for one week in a different host country. Meetings are open to any person with an interest in the subject area. The objectives of the group are communication of ideas, knowledge and experience concerning the application of Operational Research (OR) approaches and methods to problems in the health services area, mutual support between members, co-operation on joint projects, and assistance with regard to approaches and attitudes in this field.

The 31st meeting of ORAHS took place from 31st July to 5th August 2005 at the University of Southampton, UK. The theme of the meeting was "OR for Health Policy: Making Better Decisions". A unique feature of the Southampton meeting was that it ran in parallel with the second week of a EURO Summer Institute (ESI) on "OR in Health". 48 ORAHS delegates joined the 35 ESI participants who attended the first week. These 83 scientific participants were from an impressive 21 different countries. A total of 41 scientific papers were presented at ORAHS, some of which were submitted for a Special Issue of the Journal of the OR Society. This special issue is scheduled to appear in early 2007. All ESI participants are now members of ORAHS and eleven of the ESI papers will appear in a special issue of the European Journal of OR (EJOR) during 2007.

The conference had five keynote speakers: Geoff Royston (Head of OR, UK Department of Health), David Paynton (Chair of the Southampton City Primary Care Trust), Simon Dodds (a Consultant Vascular Surgeon from Good Hope Hospital NHS Trust), David Bensley (Operational Research Programme Manager, UK Department of Health) and Martin Pitt (Peninsula Medical School, Universities of Exeter and Plymouth). There was a further 14 sessions, including sessions on *Capacity Planning*, *Access and Service Quality*, *Community Care*, *Optimising the Use of Facilities*, *Emergency Department Models* and *Disease Models*. The conference also included a study visit to the Royal Hampshire County Hospital (RHCH) in the historic city of Winchester. RHCH is a UK National Health Service (NHS) Hospital and during the visit delegates had an opportunity to see examples of current innovative NHS services, including the construction of a dedicated elective treatment centre.

The two weeks (ESI and ORAHS) was totally exhausting but the most tremendous fun. We would like to thank the organizing committee as well as Charlotte Rugby from the School of Management, whose organisational skills and hard work behind the scenes helped make the two week marathon go amazingly smoothly. Many thanks also go to Sigrid Shahani, who organised and coordinated the Accompanying Persons programme. The strong contingent of Southampton health OR PhD students also provided much valuable help with the social events and countless other odd jobs – many thanks to them all. Finally many thanks to Cherry Scott for her valuable assistance in editing these Proceedings.

Editors

Sally Brailsford
School of Management
University of Southampton, UK.
SO17 1BJ
E-mail: scb@soton.ac.uk

Paul Harper
School of Mathematics
University of Southampton UK.
SO17 1BJ
E-mail: prh@soton.ac.uk

CONTENTS

HEALTH SERVICE PLANNING MODELS

OPERATIONAL RESEARCH MODELLING TO SUPPORT HEALTHCARE OPERATIONS MANAGEMENT

DAVE WORTHINGTON
Department of Management Science,
Lancaster University Management School, Lancaster, UK.

Abstract

One often proffered guide for OR model building is that the model should be made as simple as possible subject to meeting its purpose.

In healthcare operations management the purpose of an OR model is often to persuade the appropriate stakeholders to modify a system in a way that (hopefully) turns out to be an improvement. Hence the choice of model can be very context-specific, depending very much who needs to be convinced of what.

This general issue is examined via a brief description of five different modelling approaches that have been used to support operations management of outpatient clinics. The extent to which the approaches might be 'fit for purpose' in different circumstances is discussed.

Keywords: model choice; healthcare operations management; appointment systems.

Introduction

This paper is written for healthcare OR practitioners and researchers (as might attend ORAHS31). It aims to address the important general issue of model choice in healthcare modelling (and perhaps beyond). It is well recognised that different aspects of healthcare activity often require very different modelling approaches, see for example Vissers & Beech (2005), which includes descriptions of modelling approaches for:

- hospital production and capacity planning;
- planning capacity of a new hospital unit;
- master scheduling of medical specialists;
- business planning for a surgical specialty;
- scheduling outpatient clinics;

- hospital admission planning and patient mix optimisation;
- balance of care for elderly patients.

However it is perhaps less well recognised that choices of modelling approaches exist *within* healthcare activity areas. The point of this paper is to highlight and address this issue by demonstrating a clear range of choices *within* one particular well-known healthcare operations management problem area. This device is used to prompt discussion of the important factors, other than the healthcare activity being studied, that might well influence model choice in practice.

Outpatient Clinic Appointment Scheduling.

The area of healthcare activity selected for this paper is Outpatient Clinic Appointment Scheduling. The problem of outpatient and general practice appointment systems will be well known to many healthcare OR practitioners and researchers from accounts in the literature, involvement in studies or indeed from personal experience. The fundamental problem is the case where patients are given appointments to see one or more doctors. The patients may arrive on time, early, late, or not at all and are seen by a doctor in order of arrival, or in order of appointment. The two primary performance measures are patients' waiting time and doctor's idle time.

Problems were recognised in health care appointment systems as early as a century ago. In 1908 when opening a new outpatient department at Cardiff Infirmary, Osler said:

> *As one who has had a long hospital experience, may I mention one essential virtue for the members of the outpatients department staff to cultivate – namely, punctuality. It is not, of course, always possible, but it is remarkable how greatly it facilitates the work of an institution when men put in an appearance at the stroke of the hour.*

Figure 1 shows the archetypal performance of an outpatient clinic. The situation depicted here is that patients are given appointments more frequently than they can, on average, be seen by the doctor. This is a very common occurrence in a health system which wants to avoid the medical staff being idle because patients do not attend (DNA), or because consultations for some patients happen to be shorter than average. As can be seen from the Prob(busy) line, the doctor is almost certain to be busy throughout the clinic, except during the planned 10

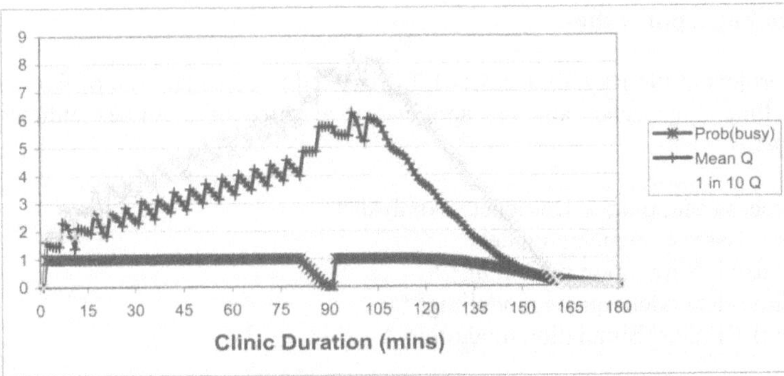

Figure 1: Archetypal performance of an outpatient clinic

minute break 80 minutes into the clinic, until the point is reached when the last patient has arrived. At this point the doctor will have seen all of his/her patients before becoming idle. The unfortunate consequence of this virtual guarantee that the doctor never has to wait, is that there are often a substantial number of patients waiting to see the doctor. In this case, on an average day, the number would typically grow from 1 to 6; and for a 1 in 10 busy clinic the number would typically grow from 2 to 8.

In 2003 a specific example of the outpatient clinic problem was faced by the Ophthalmology Department at the Royal Lancaster Infirmary, a major hospital in the north of England. It operated five or six clinics per week, involving two Consultants, three Junior Doctors and a Senior House Officer (SHO). The numbers of patients booked in for the clinics varied greatly, with some sessions finishing early, resulting in a considerable amount of 'lost' clinician time, and others finishing late, leading to over-running clinics and patients appointments being cancelled at the last minute. As can be expected, waiting times at the clinics also varied greatly. It was not uncommon for patients to be waiting for well over an hour for a consultation that could take as little as five minutes.

A study was undertaken to investigate the potential for re-organising the clinics so that more patients could be seen than at present. Although reducing patient waiting times at the clinics was not an explicit aim of the project, it was envisaged that an improved clinic schedule would also help reduce waiting times, or at least distribute them more equitably. Readers wanting more details of the study undertaken and the models used can refer to Worthington, Goulsbra & Rankin (2005) and Goulsbra & Rankin (2003).

Modelling Approaches

The project at the Royal Lancaster Infirmary led to consideration of five distinct modelling approaches and the application of three of them (as indicated ** below).

1. Process Mapping & Bottleneck Analysis**
2. Steady-state queueing models
3. 'Partial' Simulation modelling**
4. Time-dependent queue modelling**
5. 'Full'/'Fuller' Simulation modelling

1. Process mapping & bottleneck analysis

Process mapping and bottleneck analysis are well-established operations management tools, and their use has been recommended by the NHS Modernisation Agency via their series of six Improvement Leaders' Guides to encourage and facilitate this process. (The NHS Modernisation Agency was established in 2001 to support the NHS in England in the task of modernising services and improving experiences and outcomes for patients. It was superseded in 2005 by the NHS Institute for Innovation and Improvement. (See NHS Modernisation Agency (2006) for further information.) The guides cover a series of topics, establishing a framework for 'improvement leaders' to follow. The first three guides in particular look at how departments can identify areas for change. The topics covered are Process Mapping, Analysis and Redesign (NHS Modernisation Agency, 2002a); Measurement for Improvement (NHS Modernisation Agency, 2002b); Matching Capacity and Demand (NHS Modernisation Agency, 2002c).

Process mapping helps the operations manager and stakeholders to see and better understand the important features and stages of a perceived problem situation, and bottleneck analysis uses simple formulae to help identify problematic stages in the process. For example, in the Royal Lancaster Infirmary study simple formulae such as:

$$\text{Clinic Capacity} = \frac{\text{Clinic duration}}{\text{doctors} \times \text{mean consultation time}}$$

were used to give a clear indication of which of the clinic resources were likely to be a major constraint. In this case the bottleneck was clearly the consultation with the doctor.

In general terms processing capacity of the doctors can be increased if either utilisation of doctors' time is increased, or consultation times are reduced. In this case-study the interest was in the former rather than the latter. Discussion with those involved, observation of clinics and analysis of available data was therefore undertaken to ascertain what scope there was for increasing utilisation. This approach eventually focused on clinic durations, patient attendance (and non-attendance) and appointment systems:

- were too many clinics starting later than planned and/or finishing earlier than planned?
- could patient non-attendance for their appointments be reduced?
- how could revised appointment systems help achieve these aims?

In general this type of modelling enables stakeholders to appreciate the nature of the problem that they are facing, and to identify the type of solution that could improve the situation. However it gives no guidance on the precise form of the solution, or on the scale of change in performance that can be expected. In addition a significant danger with this approach is that it tends to be rather naïve, for example as one bottleneck is removed another may be caused somewhere else in the system.

2. Steady-state queueing models

It is very rare to apply steady-state queueing models in the context of outpatient clinics these days. However early work using this type of model provided important generic insights for outpatient clinic management. In common with the bottleneck analysis, steady-state queueing models reinforce the message that improvements in waiting times will be achieved by:

- reducing consultation times
- increasing the number of doctors
- modifying the arrival rate of patients to better match clinic throughput.

In addition these queueing models also take specific account of the important effects of random variation and the need to reduce it as far as possible. For example they show that, typically, an appointment system is better than no appointment system, it is desirable to match appointment times to expected duration of consultations, and it is desirable to reduce the number of DNAs where possible, particularly early in a clinic. Insofar as random variation cannot be completely controlled, Jackson (1964) also recommended traffic intensities for clinics of between 0.85 and 0.95.

However, whilst steady-state queueing models can be used to give valuable insights into the general effects of certain changes to appointment systems, consultation durations and numbers of doctors, they are not capable of making accurate predictions because clinic behaviour is typically very time-dependent.

3. 'Partial' simulation modelling

Given the stochastic nature of the problem, simulation can, in principle, be used to improve understanding of system behaviour and to predict the impact of alternative appointment systems. The term 'partial' simulation is used here to contrast with the more ambitious use of simulation, which typically endeavours to model the 'full' system incorporating all the important stochastic aspects of the problem. A 'partial' simulation focuses on a particular aspect of the problem, knowingly ignoring other stochastic aspects, but nevertheless aiming to improve understanding.

In this Royal Lancaster Infirmary case-study partial simulation was used to address two different issues, namely to discover the practical clinic capacity and to test a proposed revised appointment system.

Clinic capacities

A simple spreadsheet programme to simulate the sums of randomly generated consultation times (& DNAs) shows that the simple bottleneck equation is inadequate when trying to calculate an actual clinic capacity, even after some allowance has been made for an average DNA rate. In particular:

$$\text{Clinic Capacity} \neq \frac{\text{Clinic duration}}{\text{doctors} \times \text{mean consultation time}}$$

$$\text{Clinic Capacity} \neq \frac{\text{Clinic duration}}{\text{doctors} \times \text{mean consultation time} \times (1 - \text{DNA rate})}$$

The results of this 'partial' simulation of a 12-patient consultant clinic are shown in Figure 2, which highlights the probability distribution of clinic duration (assuming the doctor never has to wait for a patient). Clearly the average duration will be about 165 minutes, but there is a 1 in 20 chance that it could be as short as 130 minutes and a 1 in 20 chance that it would be as long as 205 minutes.

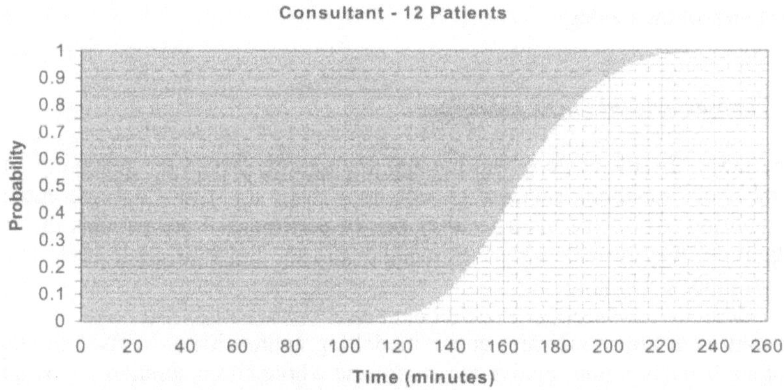

Figure 2: Cumulative probability distribution of (12-patient) consultant clinic duration.

Testing alternative appointment systems

Whilst the analysis above was very effective in highlighting the impossibility of arranging for clinics to finish at the same time each week, it was not seen as a sufficiently accurate model for testing out alternative appointment systems. A different 'partial' simulation was therefore carried out, in which five sets of real patients with randomly generated DNA's and consultation times were, manually, put through current and proposed appointment systems. The results from this analysis are summarised in Table 1.

	Current System	**Proposed System**	**Change**
Average wait	16 mins	11 mins	–5 mins
No. waiting 31–45 mins	2.8 patients	0.6	–2.2
No. waiting 46–60 mins	0.4 patients	0	–0.4
Clinic duration	2hrs 55 minutes	2hrs 56 minutes	+1 min

Table 1: 'Partial' simulation results for current and proposed appointment systems.

Whilst five simulations are clearly not sufficient to claim a high degree of accuracy, the results were sufficient to demonstrate the nature of the overall impact of moving to a different appointment system. In particular, patients would wait less long on average, with only a small proportion of patients waiting over 30 minutes. One added benefit of the manual simulations was the transparency of the model, in that clinic staff could easily be talked through each step of the model.

4. Time-dependent queue modelling

One major disadvantage of the manual simulations is that the relatively small number of randomly sampled patients may make the results unrepresentative. In addition, whilst the average measures of performance are perhaps the most important, they provide very little insight into the range of clinic performance levels to expect in practice.

In contrast a time-dependent queue modelling approach can be used to ensure that statistically representative results for the whole clinic duration are obtained very efficiently.

This modelling approach was implemented using CLINIQUE, software, which allows the user to investigate the impact of different appointment schedules on numbers of patients waiting at different times in the clinic and their associated waiting times. CLINIQUE is purpose-built software for modelling appointment systems, and is based on the queue modelling research of Brahimi and Worthington (1991) and Wall and Worthington (1999). It incorporates statistical information on patients' consultation times and overall DNA rates to predict system performance measures for appointment systems specified by the user.

Figure 3 shows that in an 'average' clinic under the existing appointment system the number of patients in the queue (i.e. in the waiting room) will range between 1 and 3 for much of the clinic, and so patients will experience waits between about 6 and 18 minutes (as consultations average about 6 minutes). Under the proposed appointment system the 'average' clinic will have considerably shorter queues that range between 0 and 1 for much of the clinic, i.e. waits between 0 and 6 minutes. In addition, as not all clinics are average, Figure 3 also shows the estimated queue lengths that would occur in the busiest 1 in 10 clinics. For the existing system queues range between 2 and 5 patients for much of the clinic, i.e. waits between 8 and 30 minutes; for the proposed system queues range between 0 and 2 patients for much of the clinic, i.e. waits between 0 and 8 minutes. Clearly on both the average and the 1 in 10 measures the proposed system is much better for patients than the existing system. However the model also indicates that the probability both doctors are busy is substantially lower in the proposed system, and hence that this particular proposal may have moved too far in favour of reducing patients' waiting times at the expense of under-utilising doctor time.

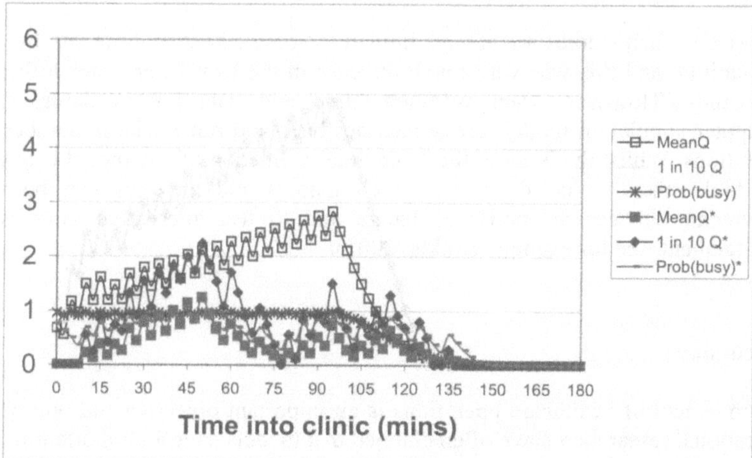

Figure 3: Performance of Registrar Clinics under previous and proposed appointment systems. (NB: * indicates proposed appointment system)

The limitation of the time-dependent queue modelling approach in some settings, is that it may not be able to model potentially important features of the system. For example it could not deal with a network of patient flows, where a patient may be delayed significantly not only before seeing a doctor, but also by waiting for other tests or treatments. At a finer level of detail it is not able to show, for example, the added benefits of scheduling patients with a higher chance of not attending later in the clinic.

Nevertheless the graph is able to show clearly and efficiently the benefits of moving to a better appointment system, not only on average, but also in terms of comparably busy days and comparably quiet days.

5. 'Full'/'Fuller' simulation modelling

In theory simulation models can always be used to model time-dependent behaviour of 'simple' and 'non-simple' clinics, as they have the capability to represent a very wide range of stochastic situations. General purpose simulation software (for example Witness and Simul8) have this capability, and specialist clinic simulation software also exists (for example CLINSIM, see Paul (1995)).

In practice such studies are always limited to some extent by time and by data availability, and this was very much the case in the Royal Lancaster Infirmary case-study. However when sufficient time and data are available, 'full' computer simulation models are sometimes built and run for large numbers of repetitions. Whilst the issue of the 'complete' validation of any model is always debatable, there is no doubt that such models and analyses can be very convincing devices to motivate change and to test alternative systems of management, see for example Dodds (2005).

Discussion

Improvement of healthcare operations is an important objective and one where operational researchers have often claimed that models have an important role to play. In theory models should enable operations managers to:

- better understand problematic situations,
- generate potential improvements,
- evaluate those potential improvements.

In practice such modelling work often needs to happen within a complex situation, where stakeholders may have competing objectives, system behaviour is not fully understood, only limited data is available, and the situation may be changing over time.

Hence, in practice, the choice of modelling approach needs to be sensitive to these practicalities. If:

- stakeholders can agree on objectives,
- the system of interest is well understood,
- data is available or can be collected on all important parameters,
- and the situation is stable,

a 'full' simulation can be a very convincing tool and have the best chance of correctly predicting the future performance of potential improvements.

However, in less ideal modelling circumstances, the operational researcher needs to be aware of alternative modelling approaches and be prepared to choose between them. In particular the modeller needs to be prepared to use judgement and experience to adopt or recommend a modelling approach that has a good chance of helping the healthcare manager to move forward efficiently in their search to improve problematic situations. This could well involve modelling using approaches selected from a wide range of possible approaches.

Conclusion

The issue of model choice is well recognised in the OR literature, and one often proffered guide for OR model building is that the model should be made as simple as possible subject to meeting its purpose.

However it is rarely clear what this means in practice. Indeed, in the 'heat of battle' there is a tendency for operational researchers to revert to their 'weapon of choice' rather than to consider the likely strengths and weaknesses of the available approaches.

This paper has sought to highlight the existence and importance of model choice, by describing five distinct modelling approaches available for one particular healthcare activity, each of which can be seen to have a role in supporting healthcare operations managers to bring about service improvements.

Acknowledgements

This paper has benefited from discussions of the presentation with the same title given during the ORAHS31 conference.

References

Brahimi M. and D.J. Worthington (1991). "Queuing models for out-patient appointment systems - a case study". *Journal of the Operational Research Society*, Vol. 42, No. 9, pp733-746.

Dodds S (2005). "Designing improved healthcare processes using discrete event simulation", The British Journal of Healthcare Computing & Information Management, Vol 22, No. 5 , pp 14-16.

Goulsbra R and J. Rankin (2003). *"A Study for the Ophthalmology Department General Review Clinics at the Royal Lancaster Infirmary."* MSc in Operational Research Dissertation, Department of Management Science, Lancaster University, UK.

Jackson R.R.P., J.D. Welch and J. Fry (1964). "Appointment systems in hospitals and general practice". *Operational Research Quarterly*, Vol. 15, No. 3, pp219-237.

NHS Modernisation Agency (2006), http://www.wise.nhs.uk/cmswise/default.htm, 24 March 2006.

NHS Modernisation Agency (2002a), Improvement Leaders' Guides (series 1, guide 1). Process mapping, analysis and redesign.

NHS Modernisation Agency (2002b), Improvement Leaders' Guides (series 1, guide 2). Measurement for improvement.

NHS Modernisation Agency (2002c), Improvement Leaders' Guides (series 1, guide 3). Matching capacity and demand.

Paul R (1995). "The CLINSIM simulation package". *OR Insight*, Vol. 8, No. 2, pp24-27.

Vissers J and Beech R (2005). "Health Operations Management", published by Routledge, September 2005.

Wall A. and D. Worthington (1999). "Using the Discrete Time Modelling Approach to Evaluate The Time-Dependent Behaviour of Queueing Systems". *Journal of Operational Research Society*, 50(8), pp777-788.

Worthington D, Goulsbra R & Rankin J (2005), 'Scheduling appointments in outpatient clinics', in 'Health Operations Management' edited by Jan Vissers and Roger Beech, published by Routledge, September 2005.

THE ORCHID PROJECT – USING OPERATIONAL RESEARCH TO BUILD PARTNERSHIPS FOR COMMUNITY HEALTH DEVELOPMENT

COLIN THUNHURST[1,4], MARY CRONIN[1,2] and MARGARET CURTIN[1,3]
[1]*Department of Epidemiology and Public Health, University College Cork*
[2]*Department of Applied Social Studies, University College Cork, Ireland*
[3]*Northside Initiative in Community Health (NCHE) Project, Cork, Ireland*
[4]*Department of Nursing, Midwifery and Healthcare, Coventry University, UK*

Abstract

The ORCHID Project is a collaborative venture between the Department of Epidemiology and Public Health at University College Cork and the Northside Initiative in Community Health (NICHE) Project, a community health development project, which serves the Knocknaheeny/Hollyhill area of North Cork. The ORCHID Project has used the problem structuring methods of *new paradigm* Operational Research, in combination with traditional analytic methods, to develop the problem analysis capacity and competences of a community health development group. This has been an essential element in a broader strategy to overcome some of the barriers that have stood in the way of partnership-building between community groups and statutory agencies within the health sector.

In this paper we briefly introduce the Primary Health Care Approach to health care development, in order to explain the critical importance attributed to the building of partnerships in health care planning. Drawing upon the experiences of developing and developed countries, we analyse the barriers that have hindered such partnership development. These experiences suggest that problem structuring methods, deployed effectively, could play a critical role in forming more constructive partnerships.

To illustrate this, we look to the experience of the ORCHID Project. Operational Research methods have been employed to facilitate analytic processes through which the NICHE Project has prepared: a) a proposal for a community-led primary care unit for the Knocknaheeny/Hollyhill area; b) a strategic plan for the next three years operation of the NICHE Project; and c) a project response to a community-based participatory planning exercise.

It is argued that OR methods can be used effectively in the institutional development of community health development groups. However, achieving a genuine partnership, aimed at improving the health status of communities living in areas of multiple disadvantage, requires statutory agencies to be equally open to new methods of analysis and to enter into new modes of discourse. The ORCHID Project will now work with key players from within statutory agencies to build a shared discourse between those agencies and community health development groups.

Keywords: problem structuring methods; community OR; partnership planning; health systems.

Partnerships in Health Care Planning – The Primary Health Care Approach

For twenty-five years health-sector development in developing countries has been moulded by the Primary Health Care Approach. Rooted in the experience of countries such as Cuba, China and Tanzania, the general formulation of the approach was the centre-piece of the Alma-Ata Conference. This historic gathering, which was sponsored jointly by the WHO and UNICEF, resulted (on 12 September 1978) in the publication of the Alma-Ata Declaration (WHO-UNICEF, 1978). Since then discussion of health planning in developing countries (and to a lesser but growing extent developed countries) has been generally predetermined by reference to the principles of the Primary Health Care Approach.

The Primary Health Care Approach, as it was initially formulated, embodied five integrated principles (Walt and Vaughan, 1981):

1. *An equitable distribution of resources* – that health services must be more equally accessible, not neglecting rural and isolated populations or peri-urban dwellers
2. *Community participation in decision making* – that active participation by the community in their own health decisions is essential
3. *Focus on preventive and promotive health services* – that preventive and promotive services rather than curative services should be the focus of health care
4. *Appropriate (not necessarily low cost) technology* – that the methods and materials used in the health system should be acceptable and relevant, appropriate technology not being synonymous with primitive or poor technology

5. *A multi-sectoral approach* – that health must be seen as only part of total care – nutrition, education, water supplies and shelter are also all essential minimum requirements to well-being

Missing from this initial formulation was a sixth principle, that of *decentralising health sector planning and service delivery*. It is possible that it was taken to be a logical derivative of the principle of community participation, given that meaningful community participation in decision-making is not possible unless the point at which decisions are made is moved closer to the point at which they impact. It is equally probable that the importance of more structural considerations only emerged as attempts were made to realise the original principles. Be that as it may, decentralisation has now effectively become a sixth principle of the Primary Health Care Approach. The mechanism advocated by the World Health Organisation for achieving decentralisation has been the development of a *district*-based model of health care delivery. The district, generally recommended by the World Health Organisation to comprise a catchment of between 100,000 and 500,000 people, is viewed as the critical management unit – large enough to provide a coherent and generally self-contained health care delivery system up to the secondary level (with appropriate referral mechanisms to tertiary facilities), and yet small enough to permit management decision-making to link into service provision and consultative processes right down to the community level.

In this decentralised model, the *district health management team* constitutes the primary focus for management development activities. {As we will see below, it has also emerged as a critical level for the potential effective deployment of *problem structuring methods*.} Although balance and composition may vary quite significantly in the implementation of the district health model from country to country (and even for larger countries from region to region), principles of collectivity in the internal functioning of the team predominate, as do principles of *partnership* in their external functioning. Horizontal partnerships, that is partnerships with other agencies whose policies and actions impact upon health status at the local level (such as education, transport, etc.), are necessary for achieving the multi-sectoral approach embodied in the fifth principle of the Primary Health Care Approach. Vertical partnerships, that is partnerships with organisations operating at the community level, are necessary for achieving the community-driven approach embodied in the second principle.

Obstacles to Building Partnerships in Developing and Developed Countries

As their importance to decentralised health care planning has been increasingly recognised, the obstacles faced in building partnerships, particularly those partnerships between statutory agencies and community organisations, have become better understood. Initially, given the earlier recognition of the importance of the Primary Health Care principles, lessons emerged from developing countries. More recently, the spotlight has turned towards developed countries. We examine these in turn.

Learning from the experience of Developing Countries

Sheikhupura is a district located close to the City of Lahore in the Pakistani Province of The Punjab. The Sheikhupura Primary Health Care Pilot Project was established as a test-bed for a range of management interventions, intended for wider implementation throughout the Province through the Second Family Health Project. These interventions were designed to establish more decentralised decision-making within the health sector and to enhance the effective involvement of local communities in decision-making processes. The Sheikhupura PHC Pilot Project ran from 1993 to 1998. The Project aimed to:

a) Reorganise and decentralise decision-making responsibility within the project area, giving greater authority to district and sub-district level officials for management and administration of government health facilities and providing improved health services to the rural population, and;
b) Establish and test a system of health committees at several sub-district levels to promote increased community involvement in planning, managing and staffing local facilities, with a view to increasing the sense of ownership and awareness amongst local people and thereby improving the quality, appropriateness and efficient utilisation of available health resources.

Accordingly the pilot project had two sets of interventions, management modification and community participation, each designed to address the respective aims. The project was *not* conceived nor designed as a partnership development project. However, in the process of implementation, significant lessons for the partnership building process emerged.

Tarin and Thunhurst (1998) have chronicled the way that the absence of an explicit agenda of partnership development served to undermine the achievement of the two discrete objectives – or, more precisely, to cause the subservience of the second objective to the first. As the powers of the district health management team were increased, in line with the attainment of the objective of decentralisation, officers within the team used their enhanced powers to attempt to exclude the local community from active involvement in the district-level planning and management processes. This impeded the achievement of the community participation objective. Viewed in hindsight, it can be seen that, in the absence of stronger commitment from the centre (clear political demonstration and direction), well-established professional demarcations, compounded by deep-seated structural and cultural beliefs and attitudes, combined to produce an inequitable power balance between centre and periphery, and translated into an inequitable power balance between professional and lay participants.

Learning from the experience of Developed Countries

For further insight into the nature of the deep-seated constraints – what organisational discourse analysts would term the *deep structures* of organisations – we turn to the experience of developed countries. This experience suggests that partnership development must be made an *active* process. This lesson comes from the analysis undertaken by the *Strategic Action Programme for Health Communities*, (SAPHC) Programme, co-ordinated by the Kings Fund within the UK. The SAPHC Programme aimed "to develop strategies that will make the statutory sector more effective in supporting community-based action for health improvement". In practice, the programme provided a comprehensive analysis of the barriers that impede partnership development at the local level (Pickin *et al*, 2002). These were subsequently portrayed in the form of a model, or a map, of these identified barriers and their interrelationships.

Put simply, this model consisted of five inter-linked *domains*: the ***community capacity to engage***; the ***skills and competencies of staff within the organisations***; the ***professional service culture***; the ***overall organisational ethos***; the ***dynamics of local and national political systems***. Each domain consisted of a number of inter-connected nodes. Each of these identified a specific impediment to community/statutory agency engagement. At the core of the model was a single node, identified as ***The Main Problem*** and specified as *No strategic approach to working with communities.* This main node provided the link between the five individual domains.

Revisiting and revising the SAPHC model, and extracting the particular features of central importance to our current focus, it may be seen to expose two major action domains and one critical set of linking nodes. The two major action domains (of those identified above) are enhancing the community's capacity to engage, more traditionally termed ***developing community capacity***, and ***addressing the organisational deep structures***. Based solely on the experience of the Sheikhupura Pilot Project, this more recent analysis might lead to a somewhat pessimistic conclusion. This is that interventions in the former domain might be rendered ineffective without completed impact of interventions in the latter domain. These by their very deep-seated nature might take a long period to achieve. However, the model also highlighted a core set of linking nodes, suggesting that more immediate feedback between the two domains is possible. These nodes focus on the nature of the problem exploration processes that statutory agencies deploy when faced with problem situations emerging from the community setting. The SAPHC model highlights the way that statutory agencies have traditionally *expropriated* problems from the community – only allowing them to define problems, rather than to be active partners in problem exploration. This limited view of community engagement is generally formed by other pre-conceptions which are also captured in SAPHC nodes: *lack of belief in community capacity to act; lack of understanding of the community's history and culture; over-simplistic approaches to community; little recognition of benefits of working with communities.*

It is therefore a reasonable premise (for which we will examine some a priori evidence below) that if a meaningfully shared process of problem exploration can be created, and facilitated through the use of the *problem structuring methods* of new paradigm OR, this may lead to the creation of new relationships between community organisations and statutory organisations. This relationship will in time be instrumental in challenging the *deep structures* of those statutory agencies. Evidence to support the validity of this premise may be drawn from both the developing and developed country context.

Problem Structuring Methods in the Process of Partnership Formation

Problem structuring methods and strengthening community capacity

The use of *problem structuring methods* (PSMs) in the development of community capacity has been well chronicled. The literature of Community OR (Midgley and Ochoa-Arias, 2004) has illustrated, over the past fifteen years, the potential of the analytic methods of OR. On the whole, these accounts have been largely descriptive in nature; evaluative accounts have been relatively few (Thunhurst and Ritchie, 1992; Rosenhead and White, 1996).

At an early stage in the development of Community OR, authors (Rosenhead J, 1986 and 1989; Jackson, 1988) argued that the key features of the problem situations confronting community groups, equated with the alternative paradigm of OR that had seen the development of PSMs. The inference was that *only* PSMs had a role to play in Community OR.

Experience has subsequently shown that this is only partly true. With an appropriate attention to issues of praxis, OR practitioners have used both PSMs and more conventional quantitative modelling techniques to aid decision-making within community groups. Thunhurst and Ritchie (1992) demonstrated the effective use of PSMs alongside the use of other more conventional OR methods in empowering community groups. Of the 26 case studies presented by Ritchie, Taket and Bryant (1994), 13 involved the use of techniques that might be categorised as problem structuring. Wong and Mingers (1994) more formally analysed a number of Community OR studies. They found that hard methods and soft methods had each been used on almost exactly the same number of occasions.

In respect of problem context, the studies reviewed by Ritchie, Taket and Bryant (1994) show perhaps surprisingly few that are taken from the health sector, and these generally relate to service planning, rather than primary health care planning. One counter-example is the work of Gibbon and Cazottes (2001) who present their use of the Health Analysis and Action Cycle in Nepal. This methodology had been developed from the Participatory Appraisal of Needs and Development of Action (PANDA) approach outlined in the writings of Taket and White (2000).

The ORCHID Project

The ORCHID Project is a collaborative venture between the Department of Epidemiology and Public Health at University College Cork and the Northside Initiative in Community Health (NICHE) Project. The NICHE Project is a community health development project, which serves the Knocknaheeny/ Hollyhill area of North Cork. The ORCHID Project builds upon previous collaboration between the two partner organisations. It was formally launched in Autumn 2004 with the assistance of a project grant from the Royal Irish Academy's Third Sector Research Programme.

The Knocknaheeny/Hollyhill area is characterised by multiple social and economic deprivations. Associated with these are health disadvantages. The NICHE project strives to address these – both directly through the services that it provides to the community and indirectly through the partnerships and alliances that it builds with statutory and other non-statutory bodies, who are also endeavouring to tackle the deeper social and economic root causes.

The ORCHID Project builds upon the understandings which have emerged from the SAPHC Project and its model of partnership building. It focuses particularly on the need to avoid the 'expropriation' of problems and problem areas by statutory agencies, and on the need to develop a shared process of problem exploration.

In order to support this shared process, the ORCHID Project has turned to the past twenty years experience of Community OR, briefly reviewed above. This, it has been argued, has the potential to provide a new partnership problem exploration technology – one which will enable all sides of a partnership to participate as equals in the shared process of problem exploration. It is this potential that is being explored by the ORCHID Project.

The Project Approach

The ORCHID Project aims to put community health development groups more firmly in the driving seat in their partnership relationships with statutory organisations. Initially, this is done by supporting independent strategic planning within the organisation. This will enable the organisation to enter the partnership process with clarity over its own objectives and confidence in its own strategy.

The first phase of engagement of the ORCHID Project has been to use the PSMs of new paradigm OR, (in combination with more traditional analytic methods), to develop the problem analysis capacity and competences of the community health development group. The Project adopts a relatively standard strategic planning approach, one which views the strategic planning process as operating as a 'planning spiral' (Green, 1992). The planning spiral represents the planning process as developing, cyclically and progressively (though not necessarily linearly), through the stages of the design, implementation and evaluation processes: Situational analysis; Priority, goal and objective setting; Option appraisal; Programming; Implementation and monitoring; and Evaluation.

Within the health field, community development initiatives have frequently focused almost exclusively on the first of the activities within the first stage of the planning cycle – the situational analysis. A range of interesting and often highly innovative techniques are available to undertake community-based health needs assessments. These tend to range from the more quantitative approaches, which use secondary data sources such as the census and locally available morbidity and mortality data, through structured qualitative approaches, which use interviews and surveys, to approaches that are derived from the more recently developed methods of rapid appraisal. The more quantitative approaches offer the virtue of 'generalisability', in that it is possible to locate communities within the broader picture. The more qualitative approaches offer the virtue of 'richness', in that it is possible to build up a fuller, more locally driven, perspective on community health needs.

The use of a plurality of approaches is preferable, though the community can often find itself confronted with information overload – picking at random from the plethora of issues that has emerged, or stumbling aimlessly from one critical campaign to another. Faced with the range of health-determining features of their environment – which will stretch from immediate health service issues, such as the availability of accident and emergency facilities, through to local environmental issues, such as housing and transport, to broad societal issues, such as poverty, discrimination and unemployment – community health development groups understandably find it difficult to focus on just one or two specific issues. [There may also be an intuitive resistance to doing this, as it may look (particularly to the group itself) as if they are ignoring other critical issues.] Where focus is achieved, this will tend to be towards the more immediate, which are often the more superficial, health determinants. This may leave the deeper underlying structural issues to the side. However, it is often the underlying determinants that may have to be addressed to achieve the more sustainable health impacts. Community health development groups face their own particular version of the upstream-downstream dilemma. How can they strike the appropriate balance between responding to more immediate needs whilst not neglecting the underlying causes?

Approaching such critical, but often seemingly intractable, strategic issues can be facilitated by adopting a more structured approach. The PSMs of new paradigm OR offer a menu of such approaches (Rosenhead and Mingers, 2001). Suitability of approach may vary with the extent to which the situation is characterized by complexity, uncertainty and conflict. Given that using such methods will, almost certainly at an initial stage, be a facilitated process, choice of approach may depend upon the style and preference of the facilitator. Where the required emphasis is one of achieving prioritisation, our preference has been to draw upon the *strategic choice approach* and upon methods derivative of it (Friend, 2001). The staged progression through *shaping*, *designing*, *comparing* and *choosing*, (Figure 1) has helped guide community health development groups from the undifferentiated assemblage of issues initially thrown up to clearly prioritised strategic options.

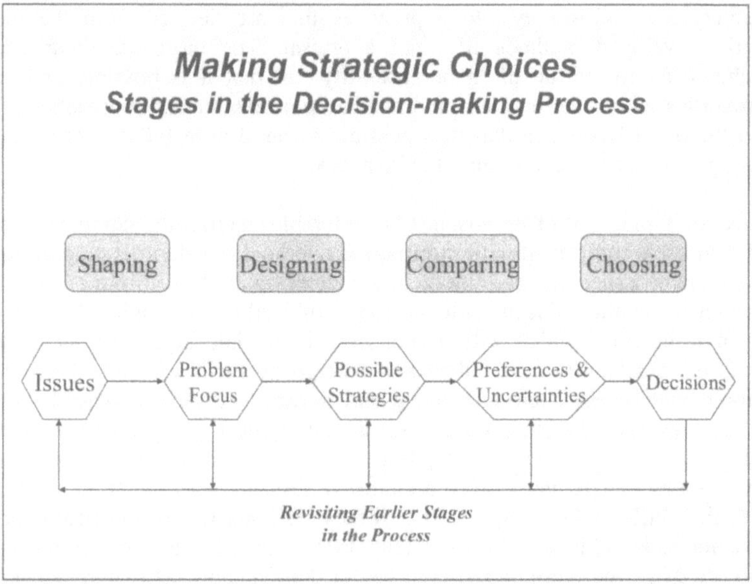

Figure 1: How the *Strategic Choice Approach* informed the planning process.

Project Impact

A community health planning exercise and the NICHE strategic plan 2004–7

The launch of the ORCHID Project followed an earlier community health planning exercise in which members of the community and of the Project had been trained (by the Talamh Nua Integrated Planning Group) in community planning techniques derivative of methods of rapid appraisal. These methods had subsequently been deployed to identify community health problems and priorities (NICHE, 2003).

The community health planning exercise had identified a broad range of issues, categorised under eight major themes. Early engagement with the ORCHID Project focused on employing methods of project planning to undertake a first prioritisation of these issues in preparation for drafting the NICHE Project's Strategic Plan for 2004-7.

Analysis of small area population statistics

Shortly following the completion of the participatory planning exercise, and the launch of the NICHE Strategic Plan, small area statistics became available from the 2002 Census of Population (which had been delayed by a year in Ireland as a consequence of the Foot and Mouth Disease outbreak). These added a significant dimension to the information base available to the NICHE Project, one which highlighted the area's particular demographic and socioeconomic features. An early activity of the ORCHID Project was to prepare an analysis of the small area population statistics drawing out these particular features (Figure 2).

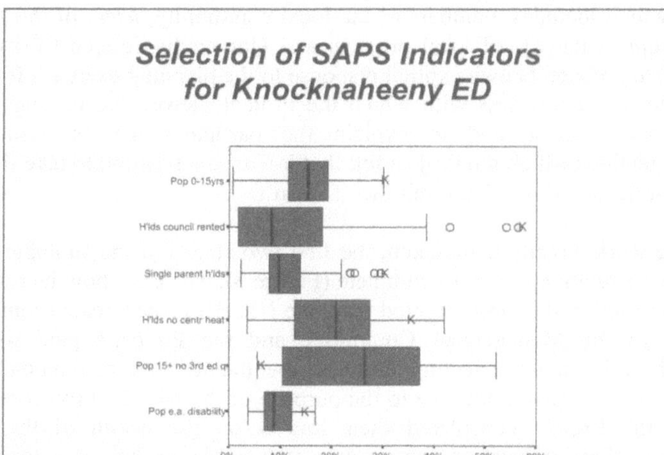

Figure 2: The analysis of small area population statistics. This diagram shows the Knocknaheeny Electoral District (ED), (marked with a K), in relation to all other EDs in Cork City. The boxes show the range within which the majority of EDs are concentrated. The vertical bar in the centre of the box shows the overall value for Cork City.

This analysis highlighted that Knocknaheeny has a young but ageing population. It has a high proportion of households with specific needs. Comparison with the preceding full (1991) census and mid-term (1996) census showed the changing nature of these needs.

Of particular note, in respect of its significance for the reproduction of social and health disadvantage, is the lag in the achievement of educational qualifications. This is particularly marked for the tertiary sector. Although the proportion of the population with a tertiary educational qualification in Knocknaheeny has increased noticeably over recent years, the increase has been less than the City-wide increase. Thus, there has been an increase in the comparative disadvantage in this critical area.

The response to the community health planning exercise

Although the community health planning exercise had been drawn upon to develop the NICHE Strategic Plan, the Management Committee of the NICHE Project (which includes members of the local community, *users* of the NICHE Project, representatives of local projects and University College Cork) felt it important to prepare its own explicit response to the planning exercise. It wished to indicate the seriousness with which the Project viewed the findings of the participatory exercise and to explain the particular themes, issues and recommendations which the Project felt that it *was* in a position to take forward, and by implication those that it felt that it wasn't.

Given the work already undertaken, the first two stages of the strategic choice process were relatively easy to complete (Figure 3). To select how its resources were to be used in the most targeted way, the NICHE Project team (comprising members of the Management Committee and the Project's paid workers) established criteria for selecting between the themes and recommendations. First, they should have relevance to the purpose of the NICHE Project. Second, although the Project considered their impact on the health of the whole community, their impact on particularly vulnerable groups was taken into account. {It was, for example, felt important to respond to the particular needs of young people in the area.} In addition to these considerations, the team prioritised those themes and recommendations that built on existing strengths (of individual workers and of ways of working). They also took particular account of the nature of the alliances and partnerships that either existed or would need to be created to take themes and recommendations forward.

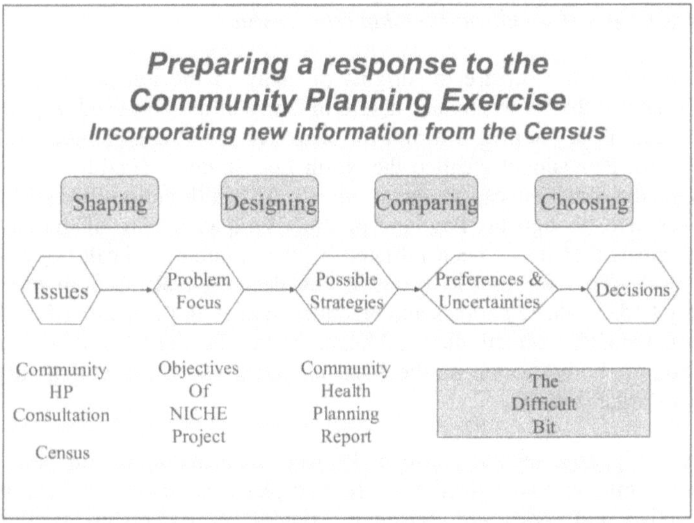

Figure 3: Informing the planning process.

Applying these criteria to the themes and recommendations, the Project team prioritized a number of areas of working. It was felt important to build on and to develop existing NICHE services (such as its drop-in sessions for men and for women, massage therapy, community walks). These were clearly much valued by the local community and their continuation was deemed important; though, as indicated already, it was felt that there was a need to target younger people, a group within the community that had been perceived to have been relatively neglected. In response to the priority accorded to improving the appearance and the safety of the locality, it was decided to initiate a local environment forum. This has drawn in a wider range of organisations and statutory bodies serving the locality. A persistent sore for the community has been the absence of primary medical facilities in the area. The one general practitioner that worked in the area had inappropriate facilities. These lacked even (at the time of the community health planning exercise) a toilet for the use of waiting patients. However, rather than limiting the perspective to that of simply improving the practice, the Project team entered into negotiation with the local health authority for the establishment of a comprehensive community-focused and community-managed primary care unit (adopting the model of the classic Peckham Health Centre of the 1930s (Scott-Samuel, 1990)).

Building deeper strategic partnership relationships

Being aware of the 'upstream' origins of many of the issues confronting the locality, and of the 'downstream' nature of many of the proposed responses, the Project team is developing a forum for proactive social and economic planning for the area. Provisionally called the North Lee Strategic Health Partnership, it builds on the experiences, (amongst others), of health action zones in the UK and elsewhere, though the key, and possibly unique, feature of the North Lee Partnership is that it has been initiated by the community health development group itself. This forum will bring in key decision-makers from the various statutory bodies whose policies and practices impact on the health of the locality – such as the City Council, the local health board, the Gardaí (police) – with the non-statutory bodies, such as the local churches and other community-based voluntary organisations.

This will trigger a second phase of Project engagement. It will be aimed at generalising the lessons gained from the first phase experience and supporting a more active and more self-conscious process of partnership building with statutory agencies.

In addition to preparing supporting analysis, as demonstrated above, local people have been trained as researchers. The ORCHID Project is currently preparing a manual, illustrated by its work with the NICHE Project, showing how the PSMs of new paradigm OR may be used alongside more traditional methods of analysis to improve the strategic planning capabilities of community health development groups.

Conclusion

The Project is an action-oriented research project using a rich case-study approach through alternative consulting. It uses methods often referred to as *systematisation* "to structure, to present coherently and to reflect analytically consulting experiences with the purpose of learning by reflecting a given praxis" (Vidal, 2002).

Initial evaluation suggests that OR methods *can* be used effectively in the institutional development of community health development groups. Community members are often very articulate, demonstrating a keen active interest; but, they do find ownership of ideas difficult to retain, and they need time to absorb and deploy OR methods.

The experience reported here is seen as a first step towards the longer term goal of developing strategic plans which are both participative and created in full partnership. Achieving this genuine partnership, one aimed at improving the health status of communities living in areas of multiple disadvantage, requires statutory agencies to be equally open to new methods of analysis and to enter into new modes of discourse. In its second phase, the ORCHID Project will also work with key players from within statutory agencies to build this shared discourse between those agencies and the community health development groups.

References

Friend J (2001). The Strategic Choice Approach. In: Rosenhead J and Mingers J, (eds). Rational Analysis for a Problematic World Revisited – Problem Structuring Methods for Complexity, Uncertainty and Conflict Wiley: Chichester, Chapter 6.

Gibbon M and Cazettes I (2001). Working with women's groups to promote health in the community using the Health Analysis and Action Cycle within Nepal. Qualitative Health Research 11: 728–750.

Green A (1992). An Introduction to Health Planning in Developing Countries. Oxford Medical Publications: Oxford.

Jackson M (1988). Some methodologies for community operational research. J Opl Res Soc 39: 715–724.

Midgley G and Ochoa-Arias A (2004). Community Operational Research – OR and Systems Thinking for Community Development. Kluwer Academic Publishers: Dordrecht, The Netherlands.

NICHE (2003). Knocknaheny/Holyhill Community Health Planning Report 2003. Northside Initiative for Community Health: Cork.

Pickin C, Popay J, Staley K, Bruce N, Jones C, and Gowman N (2002). Developing a model to enhance the capacity of statutory organisations to engage with lay communities. J Health Serv Res Policy 7: 35–52.

Ritchie C, Taket A and Bryant J (eds) (1994). Community Works. PAVIC Publications: Sheffield.

Rosenhead J (1986). Custom and practice. J Opl Res Soc 37: 335–343.

Rosenhead J (1989). Diverse unity: the principles and prospects for problem structuring methods. In: Rosenhead J (Ed) Rational Analysis for a Problematic World. Wiley: Chichester, Chapter 14.

Rosenhead J and Mingers J (eds) (2001) Rational Analysis for a Problematic World Revisited – Problem Structuring Methods for Complexity, Uncertainty and Conflict Wiley: Chichester.

Rosenhead J and White L (1996). Nuclear Fusion: Some linked case studies in Community Operational Research. J Opl Res Soc 47: 479–489.

Scott-Samuel A (ed) (1990). Total Participation, Total Health – Reinventing the Peckham Health Centre for the 1990s Scottish Academic Press.

Taket A and White L (2000). Partnership & Participation – Decision-making in the Multiagency setting. Wiley: Chichester.

Tarin EU and Thunhurst C. (1998). Community participation – with provider collaboration. World Health Forum 19: 72–75.

Thunhurst C and Ritchie C (1992). Housing in the Dearne Valley: Doing Community OR with the Thurnscoe Tenants Housing Co-operative Part 2 – An evaluation. J Opl Res Soc 43: 677–690.

Vidal R (2002). From Action to Learning – The Systematisation of Alternative Consulting Experiences, IMM Technical Report IMM-TR-2002-10, Technical University of Denmark: Lyngby, Denmark.

Walt G and Vaughan P (1981). An Introduction to the Primary Health Care Approach in Developing Countries. Ross Institute of Tropical Hygiene Publications No.13. London School of Hygiene and Tropical Medicine: London.

WHO-UNICEF (1978). International Conference on Primary Health Care. The Alma-Ata Declaration. WHO: Geneva.

Wong N and Mingers J (1994). The nature of community OR. J Opl Res Soc 45: 245–254.

MANAGING A HOSPITAL GERIATRIC WARD: IDENTIFYING COST-EFFECTIVE STRATEGIES

BARRY SHAW and ADELE H. MARSHALL
Queen's University of Belfast, Northern Ireland, UK.

Abstract

This paper presents a method for modelling the expected cost for a group of patients' accumulated time in hospital. By expanding an existing method for modelling the survival of patients in care, which incorporates a special type of Markov model in a Bayesian Network (BN), the methodology presented in this paper attaches a cost variable to the final node of the BN, enabling the estimation of departmental costs and the ability to investigate potential policy changes. An application of the model is illustrated using data relating to the admissions of 6994 patients into the geriatric department of a major U.K. hospital between 1969 and 1985. The model's practical use is demonstrated by considering the potential benefits to hospital managers investigating the implementations of making policy changes and the influence of future decisions on the hospital expenditure.

Keywords: costing; health service; hospitals, stochastic processes.

Introduction

Due to a finite supply of clinical resources, a U.K. National Health Service (NHS) budget soaring to approximately £110 billion by 2007/08 (HM Treasury, 2005) and a population becoming evermore reliant on medical care, health care managers and policy makers are experiencing increasing pressures to provide and administer their clinical resources in an optimal and cost-effective manner. Many of the methods currently in use for allocating clinical resources in hospitals are considered to have serious flaws, such as model assumptions that specify homogeneity among patients (Ivatts and Millard, 2002). For instance, the average length of stay assumes the homogeneous movement of patients through the hospital ignoring any case mix of patients that may actually exist. In order to investigate different work-efficient policies and support the hospital's operational activities, a better understanding is required of the behaviour of patients during their stay in care. This duration of stay decides the quantity of resources used by health care providers and thus has a direct impact on the expenditure of caring for the patient population.

In an attempt to more accurately model the heterogeneity of patient length of stay (LOS) in hospital, the Conditional Phase-type (C-Ph) distribution (Marshall and McClean, 2003) was developed, which incorporates a special type of Markov chain into a discrete Bayesian Network (BN) (Charniak, 1991). The Markov chain is used to model the skewed survival times of patients' length of stay (LOS). This is conditioned on the discrete component of the BN, which is used to model the relationships between the discrete influential covariates of LOS. The C-Ph distribution allows the distribution of the survival times to be estimated, with respect to the interrelationships between the causal discrete variables.

This paper extends the C-Ph model by attaching a cost variable to the LOS distribution, allowing the expected cost to be estimated for a group of patients' accumulated duration of time in care. Due to the escalating costs and restricted budgets of health care systems, such a model could be an advantage to hospital administrators and clinicians, giving them the ability to foresee the economic and non-economic consequences of any theoretical changes made to a health care system before implementing the changes on a real hospital ward.

The Model

The C-Ph distribution (Marshall and McClean, 2003) consists of two main components, the Markov model used to model the skewed survival times and the causal network of discrete variables. The Markov used is the Coxian phase-type distribution (Cox, 1955), which describes the time to absorption of a finite

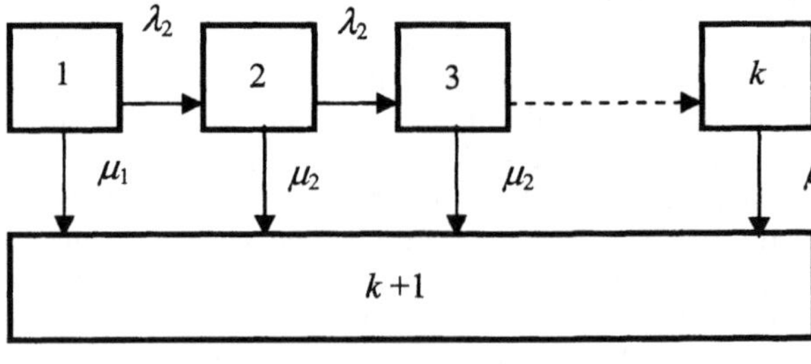

Figure 1: A k-phase Coxian phase-type distribution.

Markov chain in continuous time, where there is one single absorbing state (or phase), and the process starts in the first of k transient states, see Figure 1.

The process sequentially moves through the ordered transient states $\{1, 2, ..., k\}$, with the option of departing into the absorbing state (k+1) at any time. The parameters of the Coxian phase-type distribution, the λ_is and the μ_is, which may be estimated from the data using maximum likelihood techniques, describe the transition rates through the ordered transient states and the transition rates from the transient states to the absorbing state, respectively. The probability density function (pdf) of the Coxian phase-type distribution for time t is given by

$$f(t) = \mathbf{p}\exp\{\mathbf{Q}t\}\mathbf{q} \tag{1}$$

where $\mathbf{p}$ is a $1 \times k$ vector of probabilities defining the initial transient phases and $\mathbf{q}$ is a $k \times 1$ vector of rates from the transient states to the absorbing state. $\mathbf{Q}$ is given by

$$\mathbf{Q} = \begin{pmatrix} -(\lambda_1+\mu_1) & \lambda_1 & 0 & \cdots & 0 & 0 \\ 0 & -(\lambda_2+\mu_2) & \lambda_2 & \cdots & 0 & 0 \\ \cdot & \cdot & \cdot & & \cdot & \cdot \\ \cdot & \cdot & \cdot & & \cdot & \cdot \\ \cdot & \cdot & \cdot & & \cdot & \cdot \\ 0 & 0 & 0 & \cdots & -(\lambda_{k-1}+\mu_{k-1}) & \lambda_{k-1} \\ 0 & 0 & 0 & \cdots & 0 & -\mu_k \end{pmatrix}. \tag{2}$$

Faddy (1994) has shown that the Coxian phase-type distribution may be used to model the length of treatment times of patients in a suicide study and prisoners remand times. It has also provided a robust technique for modelling the movement of patients within a geriatric ward (Faddy and McClean 1999), the telephone service times for customers of an Israeli bank (Ishlay, 2002) and the incubation time of AIDS (Aalen, 1995). In terms of patients' duration of stay in hospital, the movements through the transient phases could represent an increasing severity of illness being treated, with the absorbing phase (k+1) representing discharge or death. A natural extension to modelling patient LOS in hospital is to consider additional variables to investigate potential influences on their LOS distribution. The C-Ph distribution extends the Coxian phase-type distribution by attaching nodes which precede it, resulting in the Coxian phase-type distribution being the last node in the BN, see Figure 2. Figure 2 consists of two components; a BN of interrelated causal nodes which precede and predetermine (in a probabilistic sense) the second component, the effect node(s) which constitute the process.

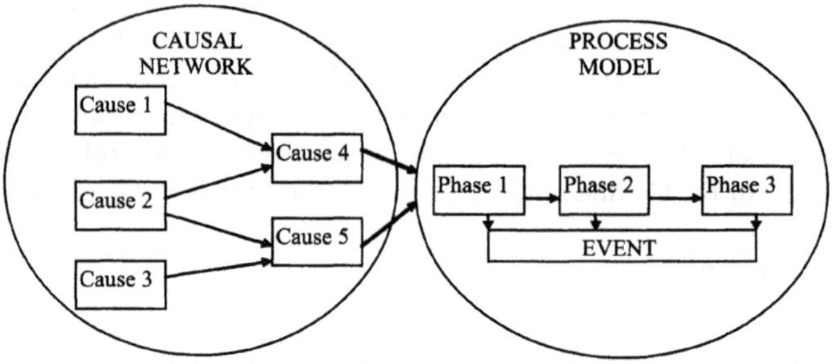

Figure 2: The Conditional Phase-type model.

The causal nodes of the C-Ph distribution $\mathbf{C} = \{C_1, .., C_m\}$ belong to the causal network and process nodes $\mathbf{Ph} = \{Ph_1, \ldots, Ph_k\}$ represent the Coxian phase-type distribution. The causal network is modelled as a discrete BN represented by the joint probability distribution

$$P(\mathbf{C}) = \prod_i P(\mathrm{C}_i \mid \mathrm{pa}\,(\mathrm{C}_i)) \tag{3}$$

where pa is the parent set of C_i. The distribution of the process nodes **Ph** in the Coxian phase-type model is then given by the pdf

$$f(t \mid \mathrm{pa(process)}) = \mathbf{p}\exp\{\mathbf{Q}t\}\mathbf{q} \tag{4}$$

where pa(process) are the parents (causal nodes) of the process and **p, Q** and **q** vary according to the values of the causal nodes in pa(process). Several algorithms (Hφjsgaard and Thiesson, 1995, Sprites and Glymour, 1991) are available in the literature for determining the best structure of the BN that most suitably describes the relationships between the variables in the data set. Subcategories of survival time data, dependent on pa(process), then have Coxian phase-type distributions fitted to them in a sequential procedure where the log-likelihood is calculated for a 1-phase distribution, and then a 2-phase distribution and so on until there is little improvement by adding an additional phase.

This paper extends the C-Ph distribution by attaching costs per bed day, which are time homogeneous but state dependent, to each of the phases of the Coxian phase-type distribution. These costs are representative of the stage of care the patient is in. Let the discrete random variable $Z^h{}_{ij}$, where $1 \le i \le k$ and $i \le j \le k$, represents the number of patients who leave hospital from phase j, given they started in phase i where h represents a combination of the possible states of pa(process). Also assume that $n^h{}_i$ represents the number of patients initially in phase i of the Coxian phase-type distribution h and $p^h{}_{ij}$ is the probability of leaving phase j, given the patient started in phase i, and thus $Z^h{}_{ij}$ is multinomially distributed. Using the joint distribution function of the multinomial distribution, the moment generating function (MGF) of the number of patients who leave the system from phase j given that they started in phase i is given by

$$M_{Z^h{}_{ij}, Z^h{}_{ij+1}, \dots, Z^h{}_{ik}}(t_{ij}, t_{ij+1}, \dots, t_{ik}) = \prod_{i=1}^{k} \left(\sum_{j=i}^{k} p^h{}_{ij} \exp(t_{ij}) \right)^{n^h{}_i} \tag{5}$$

where k is the number of phases of the Coxian phase-type distribution. As the time spent in phase i of the Coxian phase-type distribution, $T^h{}_i$, is exponentially distributed, the MGF of $T^h{}_i$ is given by

$$M_{T^h{}_i}(\theta) = \int_0^\infty \exp(\theta t^h) f_i^h(t^h) = \frac{\lambda^h{}_i + \mu^h{}_i}{\lambda^h{}_i + \mu^h{}_i - \theta} \tag{6}$$

where $f_i^h(t^h) = (\lambda^h{}_i + \mu^h{}_i)\exp(\lambda^h{}_i + \mu^h{}_i)$ is the probability density function of the length of time spent in phase i for h. Let the discrete random variable $Y^h{}_{i\tau}$ represent the cost per patient in phase i at time unit τ and $y^h{}_{i\tau}$ be some value of this random variable. Assuming the cost per patient at time unit τ is constant, c_i^h, for all τ, then the MGF of the cost per patient per time unit in phase i for h, $M^h{}_{Y_{i\tau}}(x)$, can then be given

$$M_{Y^h{}_{i\tau}}(x) = \sum_{y^h{}_{i\tau}=0}^{\infty} \exp(x y^h{}_{i\tau}) P(Y^h{}_{i\tau} = y^h{}_{i\tau}) = \exp(x c^h{}_i)\,. \tag{7}$$

If the total cost per patient (until they leave phase i) is given by the random variable, $D^h{}_i$, then

$$M_{D^h{}_i}(x) = M_{T^h{}_i}(\log M_{Y^h{}_{i\tau}}(x))\,. \tag{8}$$

The MGF of the total cost per patient, which leaves phase j, $M_{D^h{}_{ij}}(x)$, is given by

$$M_{D^h{}_{ij}}(x) = \prod_{\psi=i}^{j} \frac{\lambda^h{}_\psi + \mu^h{}_\psi}{\lambda^h{}_\psi + \mu^h{}_\psi - \theta}\,. \tag{9}$$

Assuming that the random variable T^h_N represents the total cost for all patients satisfying h for their duration of stay in the department, then using (5) and (9), the MGF of T^h_N is given by

$$M_{T^h_N}(x) = \prod_{i=1}^{k}\left(\sum_{j=i}^{k} p^h{}_{ij} M_{D^h{}_{ij}}(x)\right)^{n^h{}_i} . \tag{10}$$

The expected total cost for all patients satisfying h, $C^h=E(T^h{}_N)$, is then derived by taking the first moment of (10) and substituting $x=0$. Summing over h, the total costs of the entire group of patients in hospital is given by

$$C = \sum_h C^h . \tag{11}$$

Application

An application is considered to determine the expected costs for the accumulated duration of time in care of a group of 300 patients presently occupying beds in the department. The data set to which the methodology is applied is one referred to as the Millard data set, describing the personal and clinical details of 6994 patients admitted into the geriatric department of St. George's Hospital, London, U.K., between 1969 and 1985. The data consists of 5 variables: age ($\leq$ 85, $>$ 85 years), gender, year of admission (1969–1985), method of discharge (discharge home, transfer to another hospital or nursing home, death) and LOS (range 0–2372 days, mean 41 days, median 19 days). The large difference in the mean and median LOS indicates positively skewed data, meaning that standard statistical analysis techniques would be unsuitable for its evaluation.

Learning the C-Ph model

The software package PowerConstructor (Cheng *et al.*, 1997), which uses a 3-stage mutual information based BN construction algorithm, is used to learn the BN shown in Figure 3. The variables included in the learning process are those variables available in the Millard data set. PowerConstructor implements a constraint-based learning approach where, after ordering the nodes according to a measure of dependency between them, conditional independence tests learn the structure of the network. A GUI allows the user to specify which variables are to be analysed during the learning procedure. A node in the network represents a variable and a directed arc represents a causal relationship. A directed arc from A to B indicates that A has a causal effect on B.

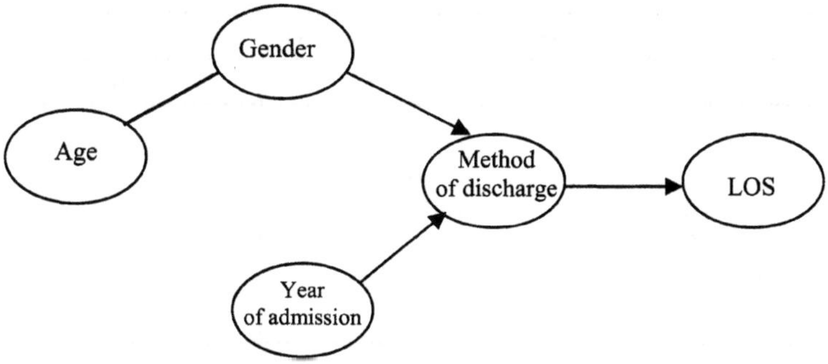

Figure 3: The C-Ph model learned from the data.

The BN identifies method of discharge as having a direct influence, or causal effect, on LOS. From a clinical point of view, this is feasible and may be explained, for example, by the fact that those patients who die whilst in hospital may be admitted in a more severe condition and hence die after a much shorter period of care than those patients eventually discharged home. This relationship agrees with other published literature, which has analysed similar clinical variables, but for different data sets (Marshall *et al.*, 2004, Shaw and Marshall, 2005). From inspection of the data values it is reasonable here to have an arc from gender to method of discharge, as males are more likely to die whilst in hospital than females. Year of admission having a causal effect on method of discharge may be due to policy changes during the 16 year period. The undirected arc between gender and age represents an association rather than causation. Again this sounds reasonable as there are generally many more older females in the observations compared to elderly males.

The Coxian phase-type distribution was fitted to the LOS data using MATLAB (MATLAB, 2004). Implementing an optimization function which uses the method of maximum likelihood estimation, MATLAB determines the log-likelihood for a 1-phase distribution, and then a 2-phase distribution and so on until the most suitable number of phases is determined using the likelihood ratio statistic. The corresponding parameter estimates of the distribution are also obtained. Table 1 displays the estimated parameters for the fitted Coxian phase-type distribution for each cohort of patients, dependent on method of discharge. Surprisingly, for each cohort the most suitable number of phases is 3. These stages of care may be, for example, interpreted as short-term, medium-term and longer-term care, or assessment, rehabilitation and long-term care.

METHOD OF DISCHARGE		
Discharge	Transfer	**Death**
μ_1=0.0018, μ_2=0.0177, μ_3=0.0044, λ_1=0.1096, λ_2=0.0005.	μ_1=0.0057, μ_2=0.0719, μ_3=0.0111, λ_1=0.1084, λ_2=0.0104.	μ_1=0.0847, μ_2=0.0344, μ_3=0.0019, λ_1=0.5009, λ_2=0.0023.

Table 1: Fitted parameters of each Coxian phase-type distribution.

Estimating departmental costs

In order to estimate the costs for a group of patients' duration of stay in hospital, the number of patients, or the number of beds, in each phase or stage of care, at time *t* must be calculated. In order to determine this, it was assumed that the hospital department is working to maximum capacity. This assumption has been used previously by different researchers (Harrison and Millard, 1991, Irvine and McClean, 1994) modelling patient flow in hospital and is a reasonable assumption to make due to today's demand for geriatric health care and the waiting lists which regularly occur. The proportion of the 300 beds allocated to each method of discharge is determined by observing the proportion of the 6994 patients in the data set who are allocated to each method of discharge. Table 2 displays the numbers of patients in each stage of care, where S_1 represents the first stage of care, S_2 the second stage and S_3 the third stage.

METHOD OF DISCHARGE								
Discharge			Transfer			**Death**		
S_1	S_2	S_3	S_1	S_2	S_3	S_1	S_2	S_3
56	**74**	**69**	**2**	**24**	**3**	**3**	**31**	**38**

Table 2: Number of patients in each stage of care, by method of discharge.

These numbers were determined by considering a constant size model and investigating the growth of a new department where new admissions into phase 1 only occur at time 0. For example, for those patients who were discharged home, 56 were initially in S_1, 74 in S_2 and 69 in S_3. By taking the expected

LOS, E(T), in each stage of care into consideration (see Table 3), the three phases are interpreted as follows: S_1 is short-term care, S_2 is medium-term care and S_3 is longer-term care. Using expert opinion, the cost per bed day (CPBD) for each of the phases of the Coxian phase-type model for each cohort of patient is estimated and displayed in Table 3. For example, for those patients discharged home, the cost per bed day for short-term care is £147, for medium-term care, £136 and for longer-term care, £145. Differentiating (10) with respect to x and substituting x=0 with the parameters from Tables 1, 2 and 3, the expected cost for each cohort of patients, C^h, can be derived where h = {discharge, transfer, death}, see Table 4.

	METHOD OF DISCHARGE								
	Discharge			Transfer			**Death**		
CPBD (£)	147	136	145	147	136	38	221	221	**160**
E(T) (days)	**9**	**12**	**90**	**9**	**55**	**227**	**2**	**27**	**526**

Table 3: Expected LOS in each phase and cost per day, by method of discharge.

Method of discharge	C^h(£)	S.D.(C^h)	**75% C.I.**
Discharge	1,057,169	92,479	**872,221 - 1,242,127**
Transfer	270,028	49,757	**170,514 – 369,542**
Death	2,909,482	444,761	**2,019,960 – 3,799,004**
TOTAL, C	**4,236,679**	**586,997**	**3,062,685 – 5,410,673**

Table 4: Expected total cost C and C^h cost for each cohort of patients.

Summing over all h provides the total expected cost for all the 300 patients' accumulated survival time in care C as £4,236,679, with standard deviation £586,997. The high standard deviation value is due to the significantly large number of longer-stay patients in the data. Chebyshev's theorem is used to determine the minimum percentage of total costs lying within a certain range, and hence Table 4 also includes the 75% confidence interval (two standard deviations from the mean) of the total costs for the duration of stay in hospital for this particular group of patients. Figure 4 illustrates the distribution of the expected number of patients still in hospital t days after the start of the study

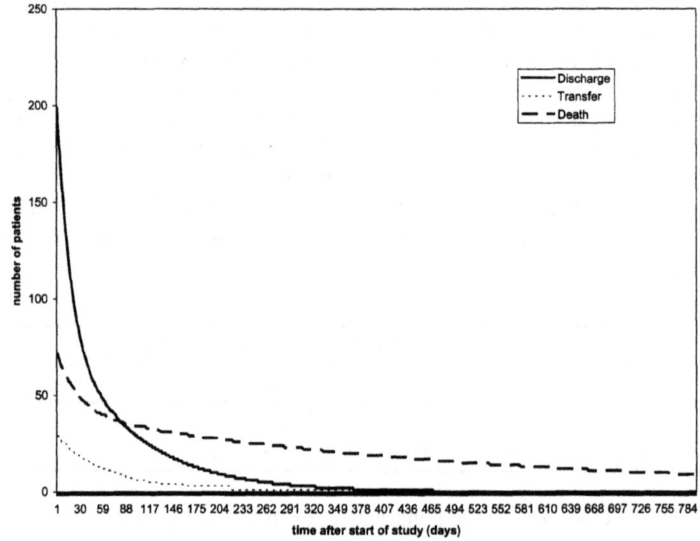

Figure 4: Expected number of patients in each phase *t* days after start of study.

period. Those patients, who are discharged home, leave hospital after a shorter period of time than those patients who are transferred or die. This is indicated by the steeper curve in Figure 4. Those patients who die whilst in hospital remain in care for a much longer period of time, indicated by the smaller gradient of their curve. To illustrate the distribution of all patients more clearly, Figure 4 has been truncated at 800 days. Table 5 displays the results of the expected accumulated cost for this cohort of 300 patients.

Year	Accumulated cost (£)	**% of *C***
1	**2,864,310**	**67.6**
2	**3,562,730**	**84.1**
3	**3,906,060**	**92.2**
4	**4,080,520**	**96.3**
5	**4,163,330**	**98.3**
6	**4,210,780**	**99.4**
7	**4,236,679**	**100.0**

Table 5: Expected accumulated cost for 300 patients.

Investigating policy changes

The continually rising cost of public health systems has resulted in methods of strategic management gaining increasing importance in health care departments. Financial mathematical models allow questions such as 'How much will it cost if...?' or 'How much could be saved if...?' to be investigated. A problem experienced by many health care providers is how to deal with the unnecessary number of elderly patients who remain in hospital for longer than they need to. A report (HCCPA, 2003) published in 2003 claimed that some hospitals were still not implementing best practice for discharging patients, and as a result there were many older people waiting for a month or more to be discharged, costing the NHS around £170m annually. The model presented in this paper now concentrates on those patients who are discharged home and illustrates how reducing the number of patients entering longer-term care affects the estimated spend-down costs. If more sufficient home-care was available for patients they could be discharged from hospital earlier instead of entering longer-term care. For instance, imagine if there was an increase in home care availability; this scenario can be considered by increasing the percentage of patients discharged from S_2, mimicking the situation of more readily available home care.

Presently, of all those patients who enter S_2, 87% leave S_2 of the hospital, with the remainder entering S_3. Keeping the expected LOS in S_2 constant and increasing the percentage leaving it to 91% results in 5 fewer patients entering S_3, a decrease in S_3 costs by 5.5% to £274,661 and a decrease in $C^{discharge}$ by 4% to £1,015,396. A further 4% increase in the percentage of patients more promptly discharged from the department is displayed in Table 6.

ALOS in S_2	12	12	**12**
% discharged from S_2	87%	91%	**95%**
S_2 costs	£231,839	£231,839	**£231,839**
Number requiring S_3 care	85	80	**75**
S_3 costs	£766,434	£724,661	**£678,711**
% reduction in S_3 costs	–	5.5	**11.4**
TOTAL, $C^{discharge}$	**£1,057,169**	**£1,015,396**	**£969,446**

Table 6: Results of increasing the percentage of patients discharged from S_2.

These results, however, assume that as the percentage of patients discharged from medium-term care decreases, the number of beds in each stage of care remains constant and the department does not regain steady-state. Such a situation would only come about if the managers enforced this through a policy change. However, as fewer patients enter longer-term care, this frees up a certain quantity of beds, which could be redistributed to other stages of care. Assuming the percentage of patients discharged from S_2 is increased to 95%, allowing the department to regain steady-state redistributes the present 56:74:69 bed allocations to a new 71:93:35. Considering a new cohort of 199 patients who are discharged home, the new bed allocation gives a new $C^{\text{discharge}}$ of £754,441, equivalent to a savings of 28%.

Conclusion

In a recent report (Wanless, 2002) it was recognised that a successful health service would not be guaranteed by higher spending alone and resources needed to be used to a maximum effect. Many methods of management used in the past and even currently, are considered insufficient. More recent methods in the literature have proposed modelling patient LOS as an indicator of resources consumed. It has been previously shown that the Conditional Phase-type distribution provides an accurate modelling technique for patients' LOS in hospital, where the different phases of the Coxian phase-type distribution represent the severity of illness of the individual and the discrete component of the BN accommodates the relationships present between the influential covariates. By attaching different costs to each of the phases, this paper has presented and illustrated a new methodology capable of modelling patients' costs and calculating the expected cost of a cohort of elderly patients for their duration of stay in hospital. By changing parameters of the model, potential policy changes can be investigated, allowing clinicians and hospital administrators the ability to foresee results before carrying out the changes on a real hospital ward. The application in this paper demonstrated that a reduction in the number of patients who are ready to be discharged home entering longer-term care, results in a saving of 28%.

References

Aalen O (1995). Phase type distributions in survival analysis, *Scandavian Journal of Statistics*, **22**: 447–463.

Charniak E (1991) Bayesian networks without tears, *AI Magazine*, **12 (4)**, 50–63.

Cheng J, Bell B and Liu W (1997). An algorithm for Bayesian belief network construction from data, *Proceedings of AI and STAT ,97*, Ft. Lauderdale, Florida, 83–90.

Cox D R (1955) A use of complex probabilities in the theory of stochastic processes, *Proc. Camb. Phil. Soc.*, **51**: 303–310.

Faddy M and McClean S I (1999). Analysing data on length of stay of hospital patients using phase-type distributions, *Applied Stochastic Models in Business and Industry*, **15**: 311–317.

Faddy M (1994). Examples of fitting structured phase-type distributions, *Applied Stochastic Models in Business and Industry*, **10**: 247–255.

Harrison G and Millard P H (1991). Balancing acute and long term care: the mathematics of throughput in departments of geriatric medicine, *Meth. Inform. Med.*, **30**: 221–228.

HCCPA (2003) House of Commons Committee of Public Accounts, *Thirty-Third Report: Ensuring the Effective Discharge of Older Patients from NHS Acute Hospitals (Hc 459).*

HM Treasury (2005). The Budget 2005.

Hjsgaard S and Thiesson B (1995). BIFROST – Block recursive models Induced From Relevant knowledge, Observations and Statistical Techniques, *Computational Statistics and Data Analysis*, **19**: 155–175.

Ishlay E (2002). Fitting phase-type distributions to data from a telephone call centre, Research Thesis: Israel Institute of Technology.

Irvine V and McClean S I (1994). Stochastic models for geriatric inpatient behaviour, *IMA Journal of Mathematics Applied in Medicine and Biology*, **11**: 207–216.

Ivatts S and Millard P H (2002). Health care modelling – why should we try?, *British Journal of Heath Care Management*, **8 (6)**: 218–222.

Marshall A H and McClean S I (2003). Conditional phase-type distributions for modelling patient length of stay in hospital, *International Transactions in Operational Research*, **10**: 565–576.

Marshall A H, McClean S I and Millard P H (2004). Addressing Bed Costs for the Elderly: A New Methodology for Modelling Patient Outcomes and Length of Stay, *Health Care Management Science*, **7**: 27–33.

MATLAB (2004). *Reference Guide*, The MathWorks Inc.

Shaw B and Marshall A H (2005). A Bayesian Approach to Modelling Inpatient Expenditure, *Proceedings of the 18th IEEE Symposium on Computer Based Medical Systems (CBMS)*, Trinity College Dublin, 491–496.

Sprites C and Glymour P (1991) An algorithm for fast recovery of sparse causal graphs, *Social Science Computer Review*, **9**: 62–72.

Wanless D (2002). Wanless Report: Securing Our Future Health: Taking a Long-Term View, Final Report.

HOSPITAL WORKFLOW MANAGEMENT: A CASE STUDY

BURCU TOKER ERSÖZ[1] and HALUK S. BALKAN[2]
[1]*Industrial Engineering Department, Girne American University, Cyprus*
[2]*Industrial Engineering Department, Eastern Mediterranean University, Cyprus*

Abstract

The study of hospital workflow has become an important aspect of the management of hospitals. This study focuses on the hospital transaction and report flow systems of Lefkoşa Dr. Burhan Nalbantoğlu State Hospital, which is located in North Cyprus. It involves the coordination and control of processes and activities of people and systems. A database model is established, and the implemented prototype workflow model system is presented. A case tool was used in analyzing the existing system at the hospital and in designing the projected system.

Keywords: hospital; database management system; workflow management.

Introduction and Literature Review

Hospital systems are among the most complex service systems. Besides giving the service of treatment, they also give the service of hospitality to patients. When a system involves mostly human beings, it becomes a very complex system.

Hospitals, which are producing health services, are defined in various ways. Some definitions of the hospital concept are as follows:

"Hospitals are places, where patients stay for control, diagnosis, treatment, and, rehabilitation, for short or long periods of time" (Mortan, 1984).

"Hospitals are organizations where the functions of patient hospitalization, diagnosis, treatment and technical services are combined. These organizations choose only one person from different disciplines" (Ak, 1990).

“Hospitals can be defined as complex and expensive hotels, which have many different and special features, and where, all the medical services are given continuously” (Ak, 1990).

In a service system, where successful treatment and hospitality are the two most important objectives, the system should be designed so that every need of the system’s users is successfully fulfilled. In this study, Lefkoşa Dr. Burhan Nalbantoğlu State Hospital, the biggest general hospital of the Turkish Republic of Northern Cyprus, is chosen as the system to be analyzed.

Since two of the other three general hospitals are weak in emergency and inpatient services, most of the patients are sent to Lefkoşa Dr. Burhan Nalbantoğlu State Hospital. As a result, the hospital is serving patients coming from all the cities and villages of the Turkish Republic of Northern Cyprus. After a number of long interviews with the experts in Lefkoşa Dr. Burhan Nalbantoğlu State Hospital, and experts in the Ministry of Health of the Turkish Republic of Northern Cyprus, the existing system is analyzed in depth.. In addition to this, major data flow and data storage problems of the hospital are detected and a solution is suggested for use by Lefkoşa Dr. Burhan Nalbantoğlu State Hospital. The systems analysis tools such as data flow diagrams, entity relationship diagrams and event response lists are used in analyzing the hospital’s system. These diagrams, except for event response lists, were prepared using the software called Visible Analyst® for system balance checking and repository forming.

After interviewing the experts and analyzing the existing system, the problems of the system are detected and solutions to the problems are suggested for a system ideally designed for the hospital. In designing the new ideal system, expert opinion, literature on hospital management and user requirements were the main sources of inputs. As a result, all the requirements and problems were taken into account when designing a new patient registration and archive system for Lefkoşa Dr. Burhan Nalbantoğlu State Hospital.

Brief Description of Lefkoşa Dr. Burhan Nalbantoğlu State Hospital

Lefkoşa Dr. Burhan Nalbantoğlu State Hospital is a well-equipped general hospital in the city of Lefkoşa in the Turkish Republic of Northern Cyprus. It serves as a second stage treatment hospital. This means that, the services are mainly provided by specialists in medical disciplines, generally requiring the use of ancillary services for the diagnosis and treatment of illness. The basic secondary care specialties are; internal medicine, surgery, paediatrics, obstetrics, gynaecology, orthopaedics, ophthalmology, urology and dermatology. It is also the only hospital, which is used half-time educational hospital that gives training to assistant doctors.

The hospital's present system is not working efficiently and does not have its own budget. There are no adequate record keeping and data acquisition systems in the hospital. This is a big deficiency and prevents the hospital keeping track of the information needed for using health systems effectively, i.e. their costs, and records of morbidity and mortality (K.K.T.C. Sağlık Bakanlığı, Sağlık Sektörü Master Planı, 2000).

Problems Related with Patient Admittance and Record System

After long interviews with the Vice Head of Doctors (Siber, R., 2000), (Bektas, S., 2000), Administrative Chiefs (Çelebi, O., 2000), Clinic Heads (Toker, A., 2000), (Refik H., 1999), (Dikengil, T., 2000) and doctors from various inpatient services (Gökçekuş, H., 1999), Head Nurse (Çoban, I., 2000), nurses from various services, and an archive officer, it became clear that there is a big problem in relation to the Inpatient Service Patient (ISP) file archive and X-Ray archive.

Although, there is an archive where ISP files and Policlinic Examination and Treatment Record (PETR) cards are stored, they are unable to find the ISP files and PETR cards when needed. The archive system of Lefkoşa Dr. Burhan Nalbantoğlu State Hospital is like a black hole. Despite the necessity of keeping one file and one protocol number for each patient, for many patients there is more than one file, file protocol number, PETR card and PETR card protocol number. Although the protocol numbers should be unique for each patient, there are many protocol numbers assigned to one patient. When a patient comes to the policlinic, or to the inpatient service, instead of finding the patient's previous medical history file, a new PETR card or file is opened and assigned a new protocol number.

This wrong way of file and record keeping has continued for many years, and now the archive system is like a bottomless well. As a result, there are long queues of patients, inquiring about their PETR cards or trying to get new PETR cards, to be seen and examined in the policlinics. In addition to this, doctors are spending their time for asking the patients about their medical histories, and all the information required to fill in the PETR cards. Also, since the new cards do not contain the patient's previous diagnosis results, treatment information, test results, and medicines used, doctors examine the patients without knowing these important details. As a result, doctors may request some of the tests again, which results in wasted time for the doctor, the patient, and the tests centers. In saving human life, loss of time is the most vital factor. This is not only loss of time, but also, the waste of resources. As the Turkish Republic of Northern Cyprus Government pays for all the health services, this is loss of money for the government.

In addition to the problem of the ISP file and card archives the X-Ray archive is a big problem because no record of the X-Rays of patients is being kept. As a result, for some patients the same X-Ray can be requested by the doctors many times, leading to long queues, loss of time and money.

The X-Ray archive problem does not conclude the problems of archive and record keeping. Another big problem related to the record keeping, is the existence of excessive quantities of documentation used for test requests and test results. These documents circulate throughout the hospital without entering the patient ISP files, so their records are not kept efficiently.

Solution of Problems Related to Patient Admittance and Record System

By drawing the data flow diagrams of the hospital's present patient information record and retrieval system, using Visible Analyst® Software, the flow of information is analyzed. In addition to the data flow diagrams, entity relationship diagrams and event response lists were also prepared. Figure 1 shows the context diagram. Figure 2 shows the data flow diagram of the hospital patient information and retrieval system. The detailed set of data flow diagrams is not included in this paper due to space limitations.

During this analysis, it was clearly observed that some information was lost during the flow in the hospital. In addition to this, some data was stored in the data stores but not retrieved. As a result, there was a loss of information. Also, it was observed that, one patient could have information distributed to many files relating to his visits to the hospital, which are present in the archive but are not retrieved for use.

After the analyzing the system of the hospital and locating the problems with the help of data flow diagrams, entity relationship diagrams and event response lists, a new system was designed with the help of the same tools for analyzing the existing system. The new patient information and retrieval system suggestion for the hospital is accompanied with an effective suggestion of change in assigning protocol numbers, which will prevent a patient having more than one file and one protocol number.

Visible Systems Corporation EDUCATIONAL/TRAINING Version

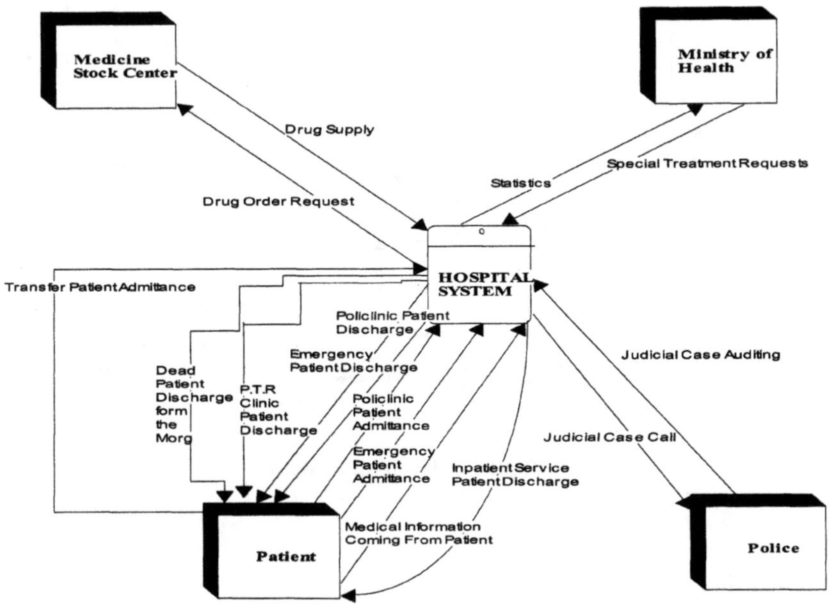

Figure 1: Context Diagram

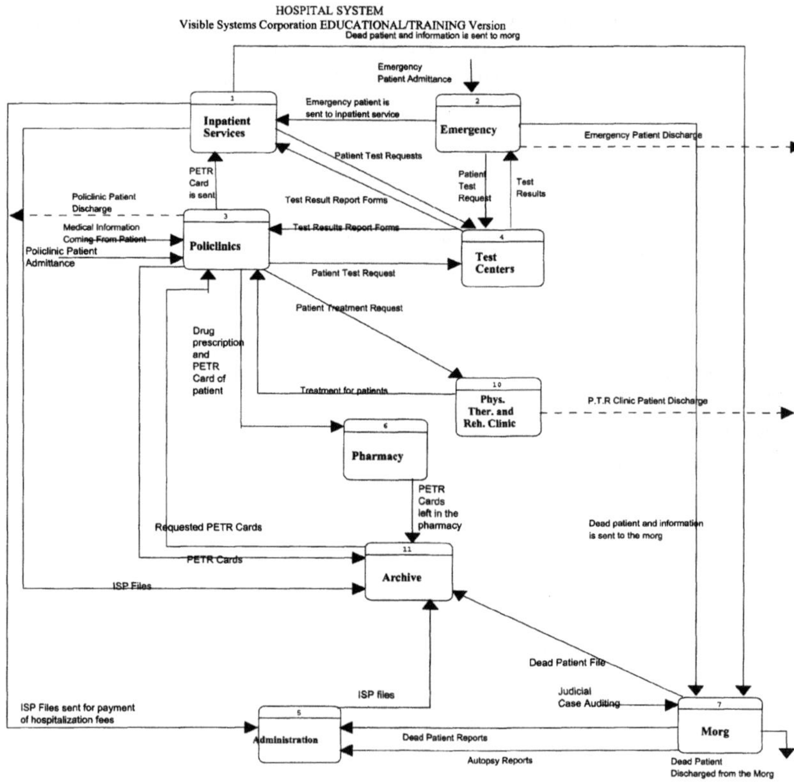

Figure 2: Hospital System

Conclusion

By analyzing the hospital's present patient information record and retrieval system, the problems were located specifically. With the help of system design tools, a new system was designed and the problems related to the patient information storage and retrieval system were solved. In the new system, instead of keeping patient information in many different files, all information related to a patient is collected in one *patient file*, which is uniquely defined by a specific protocol number. This protocol number is assigned in such a way that the patient cannot forget it and it can easily be found. This number is the patient's national identification number. Also, in the new system we suggest that, the PETR card and ISP files are combined into one patient file. In the suggested system, there will, hopefully, be no queues in the front of the policlinics, because of the efficiency of the new centralized archive system.

References

"Kıbrıs Türk Cemaati Hastaneler Yönetmeliği", Kıbrıs Cumhuriyeti Resmi Gazete, Sayı 8, 1969.

"Sağlık Sektörü Masterplanı (Redaksiyonu Yapılmamış I. Taslak Döküman)", K.K.T.C. Sağlık ve 3. Çevre Bakanlığı, 2000.

SEÇİM, Hikmet, "Hastane İşletmeciliği Seçme Yazılar", Anadolu Üniversitesi Yayınları, Eskişehir, 1995.

AK, Bilal, "Hastane Yöneticiliği", Özkan Matbaacılık Sanayi, Ankara, 1990.

Demarco, Tom, "Structured Analysis and System Specification", Yourdon Press Prentice-Hall, 1978.

Edwards, Perry, "Systems Analysis & Design", McGraw-Hill, 1993.

Martin, Merle P., "Analysis and Design of Business Information Systems" Maxwell Macmillan International Editions, 1991.

Whitten, Jeffrey L., Bentley, Lonnie D., "Systems Analysis and Design Methods", 4th Edition, Irwin McGraw-Hill, 1998.

Mortan, Kenan, "Sağlık Ekonomisi ve Eğitimi", Ankara 1984.

Cornes, Ralph, "Systems Analysis for Profitable Business Applications", Prentice Hall, 1989.

Lucas, H. C., "Information Systems Concepts for Management", 5th Edition, McGraw Hill International Editions, 1994, San Francisco.

Scriven, Donald D., Scriven Jolene D., KOZOLL Charles E., "Manual of MIS Policies & Procedures", Alexander Hamilton Institute Inc., USA.

Schultheis, Robert, Sumner, Mary, "Management Information Systems, The Manager's View", 4th Edition, McGraw Hill, 1998.

Dr. Ali Toker, Lefkoşa Dr. Burhan Nalbantoğlu Devlet Hastanesi, Fizik Tedavi ve Rehabilitasyon Bölümü Klinik Şefi (2000), Interview.

DEVELOPING AN APPLICATION OF AN ACCIDENT AND EMERGENCY PATIENT SIMULATION MODELLING USING AN INTERACTIVE FRAMEWORK

A CODRINGTON-VIRTUE[1,2], T CHAUSSALET[1],
P WHITTLESTONE[2] and J KELLY[2]
[1]*Health and Social Care Modelling Group,*
University of Westminster, London, UK
[2]*RKW Healthcare Strategist, London, UK*

Abstract

Discrete-event simulation has shown itself to be a valuable tool in modelling health care systems. This paper describes the application of a generic interactive framework to model and simulate a hospital accident and emergency department. The interactive framework facilitated user-friendly input of data such as patient pathways, arrival times and patterns, treatment times and bed resources into a discrete-event simulation software package SIMUL8. SIMUL8 was configured to enhance user friendliness and interactivity by visually showing patient flow and activity on a schematic plan of an A&E. In this paper, the generic A&E model was driven by documented or derived input including patient pathways, arrival times and patterns, treatment profile distributions and bed resources. The developed simulation model provided an insight into patient flows, patient activity, resource use and resource utilisation over the run cycle with huge potential benefit to managers and clinicians. Additionally because of its generic nature, the model could be quickly adapted to model similar A&E activities.

Keywords: simulation; OR modelling; accident and emergency.

Introduction

The potential benefits of OR modelling in health care have been identified in a number of papers including Harper and Shahani (2002), Davies and Davies (1995) and Lowery (1998). Van der Meer et al. (2005) further suggested that health care simulation models can measure both the level of services provided to patients and the efficiency with which those resources are utilised. Whilst many papers have highlighted the benefits of OR modelling, other work, for example Jun et al. (1999) and Sinreich and Marmor (2004), highlighted the poor adoption of modelling with regards to real hospitals. This paper builds upon previous work by Codrington-Virtue et al. (2005) in the development of an interactive framework for a generic accident and emergency (A&E) simulation model. The focus was to develop a generic A&E model that could be quickly and easily adapted to suit a number of different A&E operating models. The purpose of the interactive framework described in the present paper is to overcome resistances by making models user-friendly, aiding communication and participation and encouraging simulation use as a decision making tool. User-friendliness, communication and participation and simulation use as a decision making tool were identified by Barnes et al. (1997) as key elements to successful simulation in health care. The developed generic A&E model acts to manage the levels of abstraction as defined by Sinreich and Marmor, so that the complexity of highly abstracted software is shielded from the client by the user-friendly interface. This paper will discuss a practical application of the interactive framework and aims to provide an insight into levels of service and resource utilisation within an A&E setting. SIMUL8 (2004), which using Law and Kelton's (2000) definition, might be described as a stochastic, general purpose discrete-event simulation package, was used in conjunction with Microsoft Excel and Visual Basic for Applications (VBA) to develop a user-friendly interactive framework. To help validate the developed model, built around a modern service of care, model outputs were compared with real A&E data from Scotland and the hospitals supplying data.

Generic A&E Department Overview

Patient flows for the generic A&E model in this paper focus on adults and paediatric patients flowing through minor injuries (minors), rapid assessment (also referred to as major injuries unit) and resuscitation. The adult rapid assessment unit is divided further into Medicine, and Surgery and Orthopaedics. All patients had two ways into the A&E department, either by ambulance or by the walk-in (ambulant) entrance. Patients arriving by ambulance go directly to either resuscitation or rapid assessment. Walk-in patients pass through triage before treatment in either minors or rapid assessment. Figure 1 shows the generic A&E layout.

In this paper, the term 'bed' encompasses beds, bays, trolleys and cubicles and includes staff and other resources required in the treatment of A&E patients. Data to run the generic A&E model was derived from information provided by

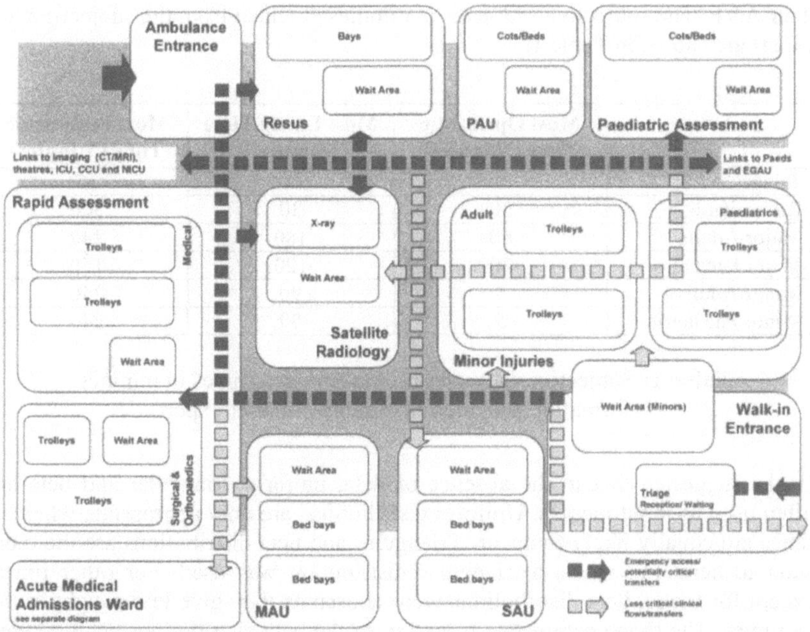

Figure 1: The generic A&E layout.

two hospitals in the United Kingdom. Figure 1 reflects a new service model layout incorporating minor, rapid assessment and resus areas. Figure 1 also shows areas assigned for short-term observation within A&E. These short-term observation areas are shown as assessment units MAU (medical), SAU (surgical) and PAU (paediatric). Figure 1 also shows separate treatment areas for paediatrics.

Process Time Data

In the absence of data and to provide a preliminary test of the generic A&E model structure, subjective estimates via subject-matter experts (as described by Law and Kelton and Banks (2001)) was the method used to determine A&E density functions. Specifically, as a subject-matter expert, a nursing manager advised subjective estimates of the most optimistic time, most pessimistic time and most likely or mode time for resuscitation, minor and major activities within their A&E. The subjective estimates in minutes obtained from the subject-matter expert are shown in Table 1.

Area	Most Optimistic Time (Minutes)	Most Likely Time (Minutes)	Most Pessimistic Time (Minutes)
Triage	5	10	15
Resuscitation	30	30	300
Major Adults	60	180	240
Major Paediatric	30	120	150
Minor Adults	5	90	240
Minor Paediatric	5	90	240

Table 1: Subjective estimates of A&E process times in minutes. Information provided by a subject-matter expert.

Banks suggested that in the absence of data, uniform, triangular and beta are often used as input models. Uniform distributions are only appropriate when all times are equally likely to occur. Triangular and beta distributions are the most suitable here. For resus, a triangular distribution was used. For other times, except for triage, beta distributions were chosen as they give less weight to the extremes. The shape parameters α_1 and α_2 of the beta distributions were derived from the subjective estimates using an approach described in Law and Kelton. With regard to triage time, a normal distribution with mean 10 minutes and standard deviation 5/3 was selected. The calculated distribution parameters for all the treatment areas are shown in Table 2. Note that the resus parameter shown in Table 2 was based on the information shown in Table 1, where parameter 'a' was the most optimistic time, parameter 'c' the most likely time and parameter 'b' the most pessimistic time.

Area	Distribution	Parameter (Minutes)
Triage	Normal	$\mu = 10.0000$, $\sigma = 1.6667$
Resuscitation	Triangular	a = 30, c = 30, b = 500
Rapid assessment unit – Medical	Beta	$\alpha_1 = 1.6667$, $\alpha_2 = 1.3333$
Rapid assessment unit – Surgery and Orthopaedics	Beta	$\alpha_1 = 1.6667$, $\alpha_2 = 1.3333$
Rapid assessment unit – Paediatrics	Beta	$\alpha_1 = 1.7500$, $\alpha_2 = 2.1875$
Minor Injuries Unit – Adult	Beta	$\alpha_1 = 1.3617$, $\alpha_2 = 1.6383$
Minor Injuries Unit – Paediatrics	Beta	$\alpha_1 = 1.3617$, $\alpha_2 = 1.6383$

Table 2: Calculated distributions.

Arrival Patterns

Patient arrival data was analysed from one of the hospitals. The arrival data recorded hourly arrivals and were collected on the first day of the month from April 2004 to March 2005. The recorded data compared visually with other A&E arrival patterns observed by Turley (1997), Coats (2001), Walley (2003) and McCaig and Burt (2004). The patient arrival times were judged to be independent and memoryless as defined by Banks. Consequently, the exponential distribution function within SIMUL8 was used in conjunction with the calculated arrivals per hour to generate arrivals within the model for both ambulance and walk-in patients. The size of the arrival data was scaled to the A&E described in the next section. Figure 2 shows comparison of the mean scaled A&E data compared with mean arrival data generated by SIMUL8. Figure 2 also shows minimum and maximum arrivals generated by SIMUL8.

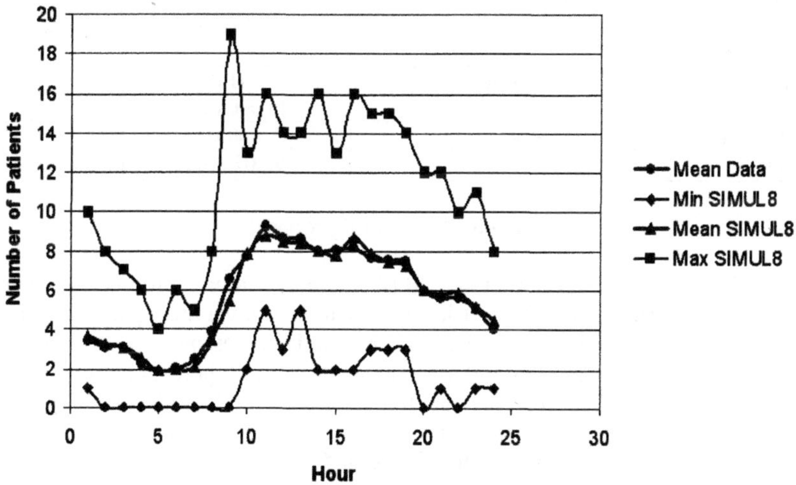

Figure 2: A&E arrival profiles.

Pathways and Bed Resources

Table 3 shows the proposed admissions for a new A&E hospital build in the UK, reflecting a new service model. Table 3 also shows the numbers of beds for the proposed A&E build. The admissions and the number of beds in Table 3, taken from the hospital build proposal, determined patient pathways through the A&E area. The pathways were determined by calculating the percentage allocated to each area in relation to the total number of admissions. The generic A&E model used beds as resources. That is to say, when patients are assigned to an area, they would consume a bed resource (with associated staff and treatment resource) on entry and release the bed resource on exit. If bed resources were not available to patients in the model, then patients would queue until bed resources were free.

Area	Admissions (Percent of Total)	Number of Beds
Resuscitation room	5,408 (8)	8
Rapid assessment unit - Medicine	12,832 (19)	18
Rapid assessment unit – Surgery and Orthopaedics	6,826 (10)	12
Rapid assessment unit - Paediatrics	3,641 (5)	6
Minor Injuries Unit - Adult	28,000 (40)	10
Minor Injuries Unit - Paediatrics	12,600 (18)	8
Total	69,307 (100)	

Table 3: A&E build proposal.

Analysis of the data shows that 40,600 (the sum of Minor Injuries Unit - Adult and Minor Injuries Unit – Paediatrics) attended minors via the walk-in entrance, whilst 28,707 attended resuscitation and rapid assessment (RA) via the ambulance entrance. Of the rapid assessment arrivals, 40 percent arrived via the walk-in entrance. Therefore, of the RA medicine total (12,832), 40 percent (5,133) arrived via walk-in. Similarly, of the RA surgical total (6,826) and RA paediatric total (3,641), 40 percent (2,730 and 1,456) respectively) arrived via the walk-in entrance. The percentage splits and routings for all ambulant and ambulance arrivals are shown in Table 4.

The asterisks (*) in Table 4 show that 19 percent of rapid assessment were routed via ambulant. Of the 9,319 total (5,133 + 2,730 + 1,456), 55 percent were directed towards RA medicine, 29 percent towards RA surgical and 16 percent towards RA paediatric. The arrival routing information shown in Table 4 fed into the pathway inputs for both ambulance and ambulant arrivals. A magnified snapshot of the walk-in (ambulant) pathways is shown in Figure 3.

Area	Admissions	Ambulance	Ambulant
Resuscitation room	5,408	5,408 (28)	0 (0)
Rapid assessment unit – Medicine	12,832	7,699 (40)	5,133 (19*)
Rapid assessment unit – Surgery and Orthopaedics	6,826	4,096 (21)	2,730 (19*)
Rapid assessment unit – Paediatrics	3,641	2,185 (11)	1,456 (19*)
Minor Injuries Unit – Adult	28,000	0 (0)	28,000 (56)
Minor Injuries Unit – Paediatrics	12,600	0 (0)	12,600 (25)
Total	69,307	19,388 (100)	49,919 (100)

Table 4: A&E arrival routing (percentages shown in brackets).

Figure 3 shows a magnified snapshot of the user-friendly interface designed to input ambulant pathway routing data. The interface here defines the pathways routings of the ambulant patients and directs the route of travel through the A&E model. For example, 56 entered adjacent to Minors (Adult) (1) directs 56 percent of the ambulant arrivals towards minor adults. Whilst 25 entered adjacent to Minors (Paeds) (2) directs 25 percent of ambulant arrivals towards minor paediatric. A further 19 percent (not shown in Figure 3 for clarity) was redirected and processed as ambulance arrivals. Figure 3 also showed that on completion of both minor injuries all patients (100 percent) were discharged. Process times, arrival patterns and numbers of beds had similar user-friendly interfaces to input data into the SIMUL8 simulation model.

Simulation Model

The SIMUL8 simulation model was configured to read inputs from the user-friendly inputs (Excel spreadsheets) to simulate patient activity through the A&E model. The simulation overlaid the generic A&E layout. In this way, patient icons activity is observable through the generic A&E layout (shown in Figure 1) in simulation time. Figure 4 shows a magnified snapshot of patient icons activity in minor injuries and triage. During the simulation run of the generic A&E model, patient icons would travel through and exit the A&E according to their assigned pathway and timings. Initially, simplified pathways were used – arrival, through treatment areas then exit. Areas not modelled

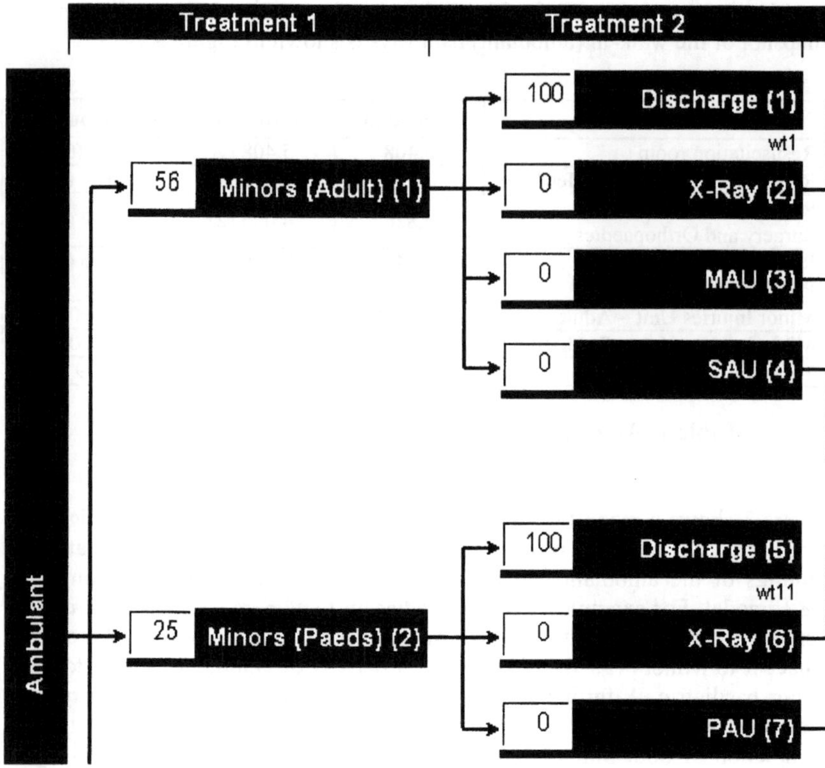

Figure 3: Walk-in (ambulant) pathway.

included X-Ray and assessment areas. The generic A&E model ran for 1 day after a warm up day. On the hour, the model captured the numbers in queues awaiting treatment. In addition, at the end of each run, data collected included the average queue times for treatment, average time from arrival to discharge and percentage utilisation of treatment areas. The trial collection period was 50 runs. The generic A&E model was validated against real A&E activity data from one of the hospital data sources and Scottish A&E data (Murphy (2005)). Although Scotland does not have a 4-hour A&E target, unlike England, the Scottish data acted as a good crosscheck for the generic model as both England and Scotland come under the NHS umbrella and shared similar working practices. Furthermore, doctors all over the UK follow similar training.

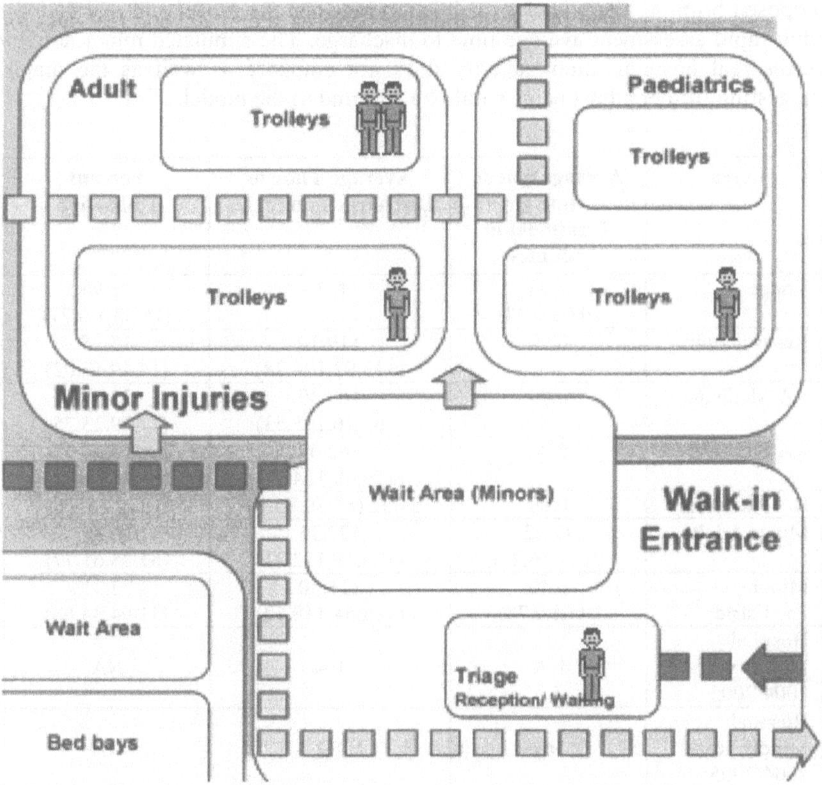

Figure 4: SIMUL8 model overlaying the generic A&E layout.

Results

Table 5 shows the average queue times for treatment, average time to discharge and percentage utilisation results generated by the generic A&E model for triage, resuscitation, RA medicine, RA surgical, RA paediatrics, minor adults and minor paediatrics. Table 5 also shows real hospital major and minor A&E activity for the year 2004/2005. Table 5 shows a close match between adult rapid assessment average time to discharge (163.40 and 162.09 minutes) and real hospital recorded data for major A&E attendance (164 minutes). On the assumption that similar treatment practices are employed in the current and the proposed hospital, there is close validation between the model and real data for adult rapid assessment average time to discharge. The simulated minor activity verses real hospital minor activity does not compare as well as the major, suggesting further adjustments would be required to the model.

Area	Average Queue Time for Treatment in minutes	Average Time to Discharge in minutes	Percent Utilisation
Triage	0.04 (0.03,0.04)	NA	15.90 (15.53,1 6.27)
Resuscitation	0.00	119.10 (113.92,124.28)	15.75 (14.49,17.02)
RA Medicine	0.00	163.40 (161.26,165.53)	22.77 (21.79,23.75)
RA Surgical	0.00	162.09 (159.40,164.77)	17.04 (15.94,1 8.13)
RA Paediatric	0.00	84.32 (82.50,86.14)	7.18 (6.55,7.81)
Minor Adults	12.22 (8.17,16.27)	127.75 (123.19,132.31)	59.80 (57.83,61.77)
Minor Paediatric	0.40 (0.05,0.74)	116.80 (113.54,118.63)	33.31 (31.94,34.68)
Hospital Majors year 2004/2005	NA	164	NA
Hospital Minors year 2004/2005	NA	89	NA

Table 5: Average queue time for treatment, average time to discharge and percent utilisation (95 percent confidence intervals shown in brackets).

Table 6 compares the generic A&E model and the Scottish A&E data. Table 6 categorised the data under the Scottish headings of resuscitation, trolley cases and walk-in, where resuscitation met a similar category as resuscitation, trolley cases matched rapid assessment (majors) and Scottish walk-in's matched walk-in's in the generic model. Ages 15 and over were defined as adults for this paper, whilst, ages 14 and under were defined as paediatric. Table 6 shows the sample results generated by the generic A&E model in brackets. Note also that the generic A&E model does not currently distinguish between adults and paediatric resuscitation patients. As such, generic data in Table 6 assigned to resuscitation is a combination of adult and paediatric patients.

Age Group		Resuscitation time in minutes	Trolley time in minutes	Walk-in time in minutes
Age 14 and under	Median	72 (80)	91 (86)	90 (104)
	90th percentile	186 (207)	191 (128)	201 (173)
Age 15 and over	Median	100 (80)	130 (170)	106 (113)
	90th percentile	249 (207)	290 (212)	234 (205)

Table 6: Comparison of real Scottish A&E Data (2004) with generic A&E model. (Generic A&E model shown in brackets).

Table 6 shows that results generated by the generic A&E model are of the same order as the Scottish data, in that longer times in the Scottish data related to longer times in the model and visa versa, adding weight to the potential further use and development of the generic A&E model. Figure 5 shows a sample output distribution for minor adult discharge. The minor adult discharge data shown Figure 5 directly relates to the bracketed age 15 and over walk-in data shown in Table 6. Table 6 (supported by Figure 5) shows 90 percent of patients were treated within 205 minutes and thus within the 4-hour target.

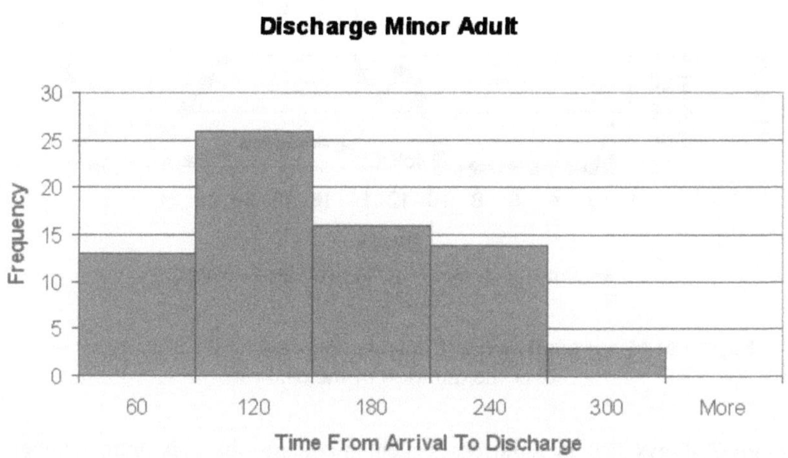

Figure 5: Histogram – Discharge Minor Adult relating to Table 6 adult (age 15 and over) walk-in.

Results generated in Table 5 showed zero or small queues and relatively low utilisation for all treatment areas except for minor adults. In contrast, minor adults generated queues and had the highest level of utilisation. In order to gain insight into utilisation levels, we focus on two main areas; minor adults with the highest queues and utilisation levels and rapid assessment medicine with its low utilisation levels yet high resource allocation (18 beds). Figure 6 shows minor adults queue information recorded on the hour, over a 50-run trial. Figure 6 shows mean queues, maximum queue and the probability of queuing over the trial period. For example, we can see that probability of queuing with this level of resource (10 beds and associated resources) rises to a peak mid afternoon, then drops away as the day progressed. In contrast, the potential queue size rises from mid morning throughout the day. It is also interesting to observe that between approximately 12:00 and 16:00 one has a chance of joining a small queue. However, as the day goes on, the chance of joining a queue falls, but the chance of joining a larger queue increases.

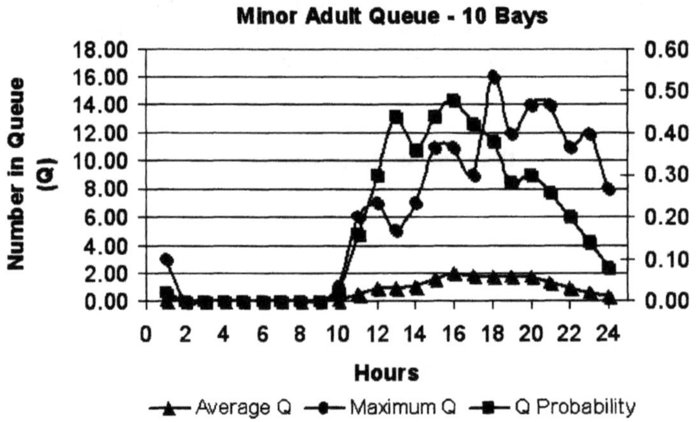

Figure 6: Minor adult queue. Contents recorded on the hour every hour over the duration of the trial run.

Figure 7 shows that in rapid assessment medicine, the maximum number of beds full at any one time was 13. This suggests that a reduction in the number of beds is possible. Re-running the model with 10 beds in the rapid assessment for medicine yields a significant increase in bed utilisation while hardly decreasing the average time to discharge – see Table 7.

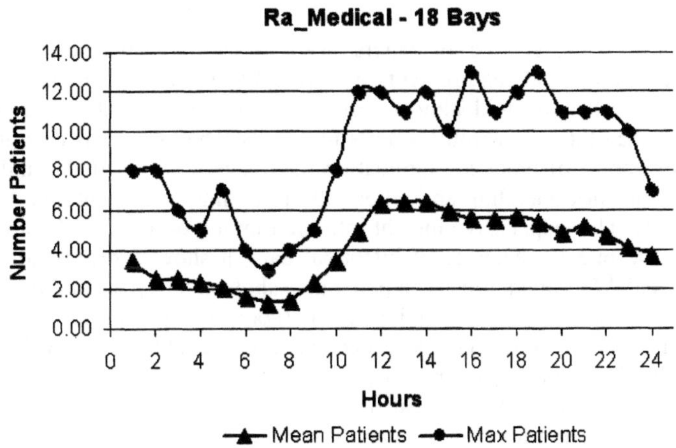

Figure 7: Rapid Assessment Medicine - contents recorded on the hour every hour over the duration of the trial run.

The model output shown in Table 7 suggests that using 18 beds, RA medical is poorly utilised. However, bed utilisation needs to be matched with the severity of potential patients and one might suggest that rapid assessment patients are treated with the shortest queues possible. Changing the number of bed resources (from 18, to 14, 10 and 6 respectively) and re-running the model, shows a clear example of the trade-off between resources use, treatment times, queuing times and how simulation can help support the management decision-making role. The trade-off in the number of bed resources is highlighted in Table 7 which shows a significant jump in the average queue time for treatment when the modelled number of RA medical bed are further reduced from 10 to 6.

	Average Queue Time for Treatment (Minutes)	**Average Time to Discharge (Minutes)**	**Percent Utilisation**
RA Medicine – 18 beds	0.00	163.40	22.77
RA Medicine – 14 beds	0.00	163.28	29.31
RA Medicine – 10 beds	0.94	164.38	40.99
RA Medicine – 6 beds	38.59	199.50	67.41

Table 7: Average queue time for treatment, average time to discharge and percent utilisation.

Conclusion and Future Work

Within an A&E area, service managers need to carefully balance service utilisation and the clinical requirements of seriously ill patients. For instance, lower levels of utilisation will almost certainly be more acceptable in resuscitation rather than the minor injuries units for adults. This paper has described and shown examples of practical application of a generic A&E model and its interactive framework operating and potentially providing insight into an A&E operational area. This paper gave insight into the running of a relatively simple A&E model. Despite the absence of historical data, it showed results not too dissimilar to real A&E operations. The paper also showed the speed and relative ease with which simulations could be re-run modifying parameters via the interactive interface. This shows that a generic model could provide valuable information relatively quickly over a variety of A&E scenarios. Further work is still required to fully validate and test the model and to further explore other scenarios such as observation wards, imaging services and other testing.

References

Banks, J., (2001). *Discrete-event system simulation.* 3rd ed. Prentice-Hall International. London.

Barnes, C. D., Quiason, J. L., Benson, C. and McGuiness, D., (1997). Success Stories in Simulation in Health Care. In: *Proceedings of the 1997 Winter Simulation Conference*, Atlanta, 7–10 December 1997, Piscataway, New Jersey: IEEE, 1280–1285.

Coats, T. J. and Michalis, S., (2001). Mathematical modelling of patient flow through an accident and emergency department. *Emergency Medicine Journal*, **18**, 190–192.

Codrington-Virtue, A., Chaussalet, T., Whittlestone, P., and Kelly. J., (2005). An Interactive Framework for Developing Simulation Models of Hospital Accident and Emergency Services. In: *Medical and Care Compunetics 2 – 114*, The Netherlands, IOS Press, 277–283.

Davies, H. T. O. and Davies, R., (1995). Simulating health systems: modelling problems and software solutions. *European Journal of Operational Research,* **87**, 35–44.

Harper, P. R., and Shahani, A. K., (2002). Modelling for the planning and management of bed capacities in hospitals. *Journal of the Operational Research Society,* **53**, 11–18.

Jun, J. B., Jacobson, S. H. and Swisher, J. R., (1999). Application of discrete-event simulation in health care clinics: A survey. *Journal of the Operational Research Society,* **50**, 109–123.

Law, A. M. and Kelton, W. D., (2000). *Simulation modeling and analysis.* 3rd ed., McGraw-Hill. New York.

Lowery, J. C., (1998). Getting Started in Simulation in Healthcare. In: *Proceedings of the 1998 Winter Simulation Conference.* Washington, DC, 13-16 December 1998, Piscataway, New Jersey: IEEE, 31–35.

McCaig, L. F. and Burt, C. W., (2004). National Hospital Ambulatory Medical Care Survey: 2002 Emergency Department Summary. *Advance Data*, **340**, 1-34.

Murphy, D., (2005). Resuscitation – Arrival to Completion. *E-mail correspondence to Codrington-Virtue, A.*, (A.Codrington-Virtue@wmin.ac.uk) 19 April *2005.*

SIMUL8 Corporation, SIMUL8 Standard R11 software (2004), http://www,simul8.com

Sinreich, D. and Marmor, Y., (2004). A Simple and Intuitve Simulation Tool for Analyzing the Performance of Emergency Departments. In: *Proceedings of the 2004 Winter Simulation Conference.* Washington, DC, 5-8 December 2004, Piscataway, New Jersey: IEEE, 1994–2002.

Turley, S., (1997). Development of the 'Euro Rota' in A & E. *Accident and Emergency Nursing*, **5**, 178–180.

Van der Meer, R. B., Rymaszewski, L. A., Findlay, H. and Curran, J., (2005). Using OR to support the development of an integrated musculo-skeletal service. *Journal of the Operational Research Society,* **56**, 162–172.

Walley, P., (2003). Designing the accident and emergency system: lessons from manufacturing. *Emergency Medicine Journal*, **20**, 126-130.

SIMULATION MODELLING OF A MAJOR ACCIDENT AND EMERGENCY DEPARTMENT IN ROME

LINDA SALUSTRI[1], PAOLA SBANDI[1], SALLY BRAILSFORD[2], ROBERTO BROCATO[3], VANDA DE ANGELIS[4], PAUL HARPER[5] and GIOVANNA JONA LASINIO[4]

[1]*Facoltà di Scienze Statistiche, Università "La Sapienza", Roma, Italy*
[2]*School of Management, University of Southampton, UK*
[3]*Azienda Ospedaliera San Camillo-Forlanini, Roma, Italy*
[4]*Dipartimento di Statistica e Statistiche Applicate, Università "La Sapienza", Roma, Italy*
[5]*School of Mathematics, University of Southampton, UK*

Abstract

A hospital Emergency Department is a complex system, characterized by high variability due to stochastic patient arrivals and the diversity of patients' medical needs. In this context, simulation can be a useful tool to understand, analyze and manage the system in order to quantify the resources required to deliver an acceptable quality of service. The simulation model described here reproduces the workings of one of the largest Accident and Emergency Departments (AED) in Rome. The simulation model was used to explore some of the major bottlenecks in the AED, which cause long waiting times and overload available resources. The model can be easily extended to study situations of exceptionally high demand, such as major incidents or disasters.

Keywords: emergency department; accident and emergency; simulation; hospital capacities.

Introduction

The Emergency Health Services (EHS) in the city of Rome are often challenged by the need to supply medical care to a high, and sometimes exceptional, number of people, while trying to reduce the impact of this unusual demand on the local population. Such surges in demand are caused by the many large events that are hosted in Rome, such as religious, socio-political, cultural and sporting events.

The EHS consists of two different but connected functions (for a full description see Gazzetta Ufficiale 1992 and 1996):

- Emergency callouts, tackled by the *Centrale operativa 118*, contacted by dialling the telephone number 118;
- Emergency health care, supplied by the Accident and Emergency Department (AED) in the hospital (Bollettino Ufficiale della Regione Lazio, 1996 and 2002).

Emergency callouts have been previously studied by De Angelis *et al* (2004), during the Jubilee Year 2000. The aim of De Angelis' study was to plan the number and location of ambulances and staff in order to guarantee that a patient reaches the nearest Emergency Department within eight minutes (Bollettino Ufficiale della Regione Lazio, 1994). The current paper considers only the AED, which has become the focus of attention due to an ever increasing routine daily demand, as well as casualties from large events hosted in the city. Furthermore, there is a need to model potential situations of exceptional demand caused by terrorist attacks.

Modelling of health care services has been the subject of many OR studies, and simulation has been successfully used in many projects concerning hospital capacities (for example Bowers *et al*, 2000; De Angelis *et al.*, 2003; Harper, 2002; Lagergren, 2004; Vissers, 1999). Indeed, AED's have themselves been the subject of many modelling studies across the world (see for example Brailsford *et al.*, 2004; Lubicz and Mielczarek, 1987; Romanin-Jacur and Facchin, 1987).

In this paper we study one of the biggest AEDs in Rome, that of the San Camillo-Forlanini Hospital, in order to understand and analyse how the department functions in a normal (routine day-to-day) situation. The prototype model presented here has assisted the hospital with the management of the AED, and provides an excellent basis for future work on modelling situations of exceptional demand.

Problem Definition

Direct observation of the processes of the San Camillo AED and several meetings with the medical and paramedical staff gave us a good understanding of the entire system. We were also able to map different patient pathways according to the urgency codes assigned to patients at triage (Gazzetta Ufficiale,

2001): red and yellow codes for major (serious) patients, green and white codes for minor patients. Through discussions with hospital managers, it was possible to identify the critical parts of the AED system on which to focus our study: namely, patient arrival rates, activity durations and resource availability. It was agreed that the objectives of our study would include the calculation of waiting times, resource utilization, queue lengths and patient time spent in the AED, the identification of serious bottlenecks in the AED and the investigation of the effects of different AED configurations.

Data Analysis

Data collection

AED activity data are managed by GIPSE (*Gestione Informazioni Pronto Soccorso ed Emergenza*), the information system of *Regione Lazio*, the area that includes the city of Rome. GIPSE data is provided at two levels: aggregated data, and individual data from patient hospital files. Due to bureaucratic problems caused by the need for patient confidentiality, and also the practical difficulties caused by the inflexibility of the software, it was only possible to obtain aggregated data from GIPSE. The AED organization had changed significantly in January 2003, making previous data irrelevant, so GIPSE data from January 2003 to June 2004 was used to provide information about the different ways of accessing the AED, waiting times, specialist examinations and diagnostic tests, bed occupancies and triage classifications. It was not possible, however, to track the complete path of each patient and this has partly restricted our study. Consequently, it has been necessary to interview doctors to gain insights and obtain estimates, for example for activity durations that were not available from GIPSE.

Data exploration

Excel spreadsheets were used to prepare and analyse the data. The first analysis was performed in order to find whether the AED arrival distribution, classified by urgency code, month, day of the week and time window of the arrivals, is uniform. We used the non-parametric χ^2 test, which showed that we only needed to divide arrivals into two distinct groups: hot months, with the highest patient arrival rate, and cold months with the lowest arrival rate (Figures 1 and 2).

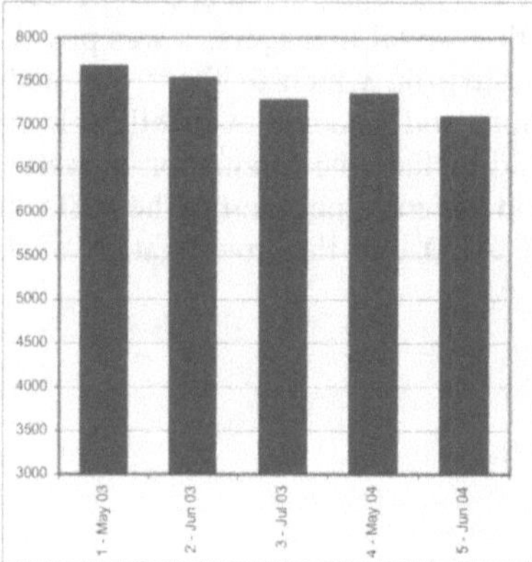

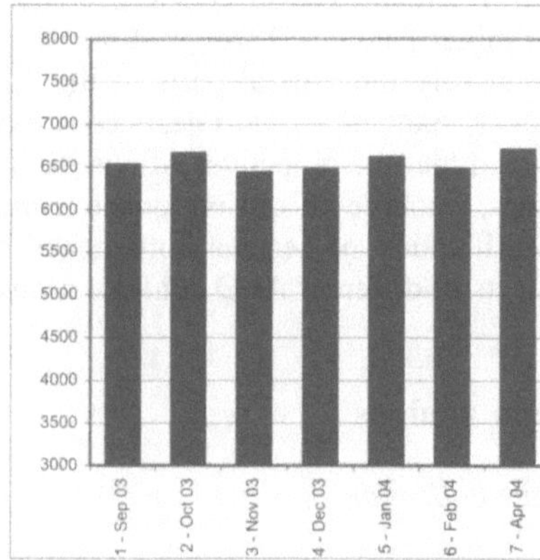

Figure 1 Arrivals in hot months. **Figure 2** Arrivals in cold months.

A second analysis verified the independence between different variables. This allowed us to show that the probabilities relating to each of the different branches of the patient pathway do not vary over time. The main branch represents arrivals split by triage codes. In fact, we found that the percentages of the four codes are the same for the various months, days of the week and time windows (Figure 3).

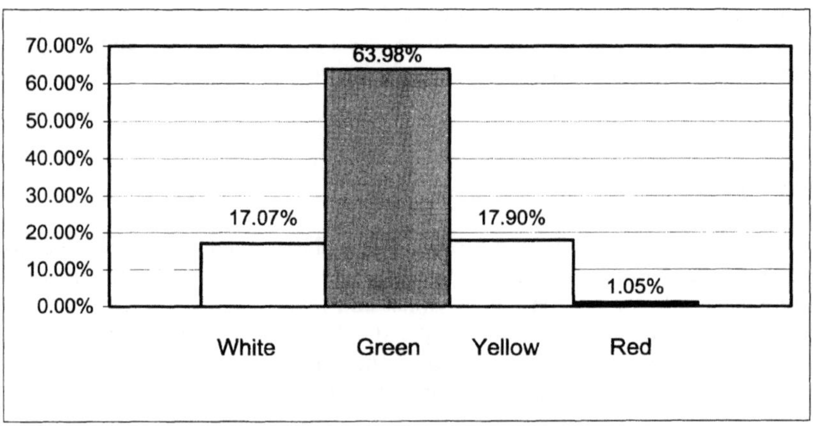

Figure 3: Probability distribution of triage codes.

We paid particular attention to the choice of the probability distributions from which to sample inter-arrival times and activity durations for the simulation. For the inter-arrival time, we chose a negative exponential distribution with mean equal to the average inter-arrival time, calculated from the available data. For the activity durations we chose the Weibull distribution, as this is very flexible in shape and allowed us to fit the parameters using expert judgment (Harper and Shahani, 2003). Lastly, to sample the variables "number of specialist examinations" and "number of diagnostic tests" we chose a Poisson distribution, with mean equal to the average number of specialist examinations or diagnostic tests for each patient, calculated from the data.

Construction of the Logical Mathematical-Statistical Model

We initially built a model to represent how the system works. The elements of the model do not correspond one-to-one with the elements of the real system, but they capture the essence of the system in sufficient detail for the purposes for which the model is intended (Law and Kelton, 1982). We needed to introduce some hypotheses in order to model the sequence of medical examinations and diagnostic tests. Firstly, when patients receive their "first medical examination", we assume that he/she first receives all the necessary specialist examinations, followed by any diagnostic tests. Afterwards, the same doctor who carried out the first medical examination examines all the test results, closes the patient's hospital file in the activity "second medical examination" and sends him/her to his/her final destination (admission to a ward, or discharge). Figure 4 shows the "life cycle" of the entities (patients) and the functional relationships between the objects in the system. It can be seen that there are different pathways for each of the different triage codes, except for white and green, since green and white code patients follow the same path and use the same resources. Red code patients are given their first medical examination immediately on arrival. Serious patients are treated in the main emergency room, while minor patients are treated either in the medical, surgical or orthopaedic examination rooms, as appropriate. According to the model logic, after the first medical examination the patient is given a label (an attribute) showing the number of specialists he/she needs to see, sampled from empirical distributions. If this number is greater than zero, the patient will, with given probability, join one of the queues for the appropriate specialist. When the specialist examinations finish (or when the label value is zero), another label showing the number of diagnostic tests to be carried out is assigned to the patient. Also in this case, he/she will stay in one of the queues for the diagnostic tests, chosen according to the probability distribution. When the patient has

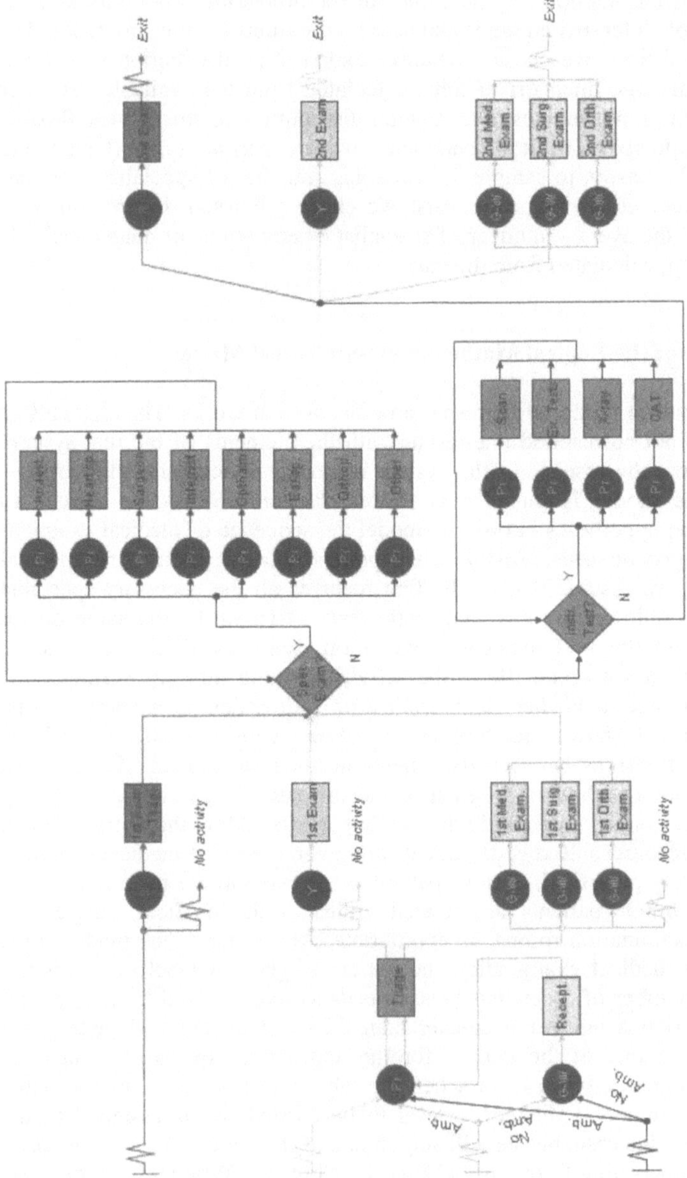

Figure 4 Activity Flow Diagram.

completed all their tests, they receive the second examination, during which the doctor decides where he/she will go from the AED (to a ward or to be discharged).

Construction of the Simulation Model

Using Figure 4 as the activity flow diagram, we built a Discrete Event Simulation model using the commercial software *Simul8*. The use of the internal programming language *Visual Logic* permitted complex patient pathway rules and logic to be implemented. In our model (Figure 5), there is only one entry point, after which the patient pathways differ according to the triage code. In the model there are a total of 60 activities, 24 of which correspond to real activities in the AED, the others being zero-time *dummy* activities used to assign labels and implement decision rules as necessary. There are also 29 different types of resource, for example doctors, nurses, technicians and beds, which can be *dedicated* to a single activity or *floating* between two or more activities. The choice of initial conditions, as well as of the length of each run and of the number of independent runs to be performed, was fundamental. Experimentation showed that a 15 day warm-up period was necessary, and this was followed by results collection for a further 30 days. Further experimentation showed that at least 100 simulation runs were necessary.

Validation, Experimentation and Results

Validation took place throughout the study. Indeed, the continuous interaction with the AED staff, the statistical tests carried out on the data and the trial runs performed during the construction of the simulation model in order to obtain the right logical flows, have all contributed to the development of a valid and complete model. Moreover, statistical analysis of the difference between a specific measure of performance of the real system (the total time spent in the system) and the corresponding value obtained from the simulation confirmed a good level of fit of the model.

We used the model to experiment with the AED system. During this phase, bottlenecks were identified and alternative scenarios formulated. We focused on the hot summer months with higher demand, thus providing the AED managers with guidance on the resource needed to cope with the busiest times. The analyses have concerned not only the total time spent in the system, but also resource utilization and patient waiting times.

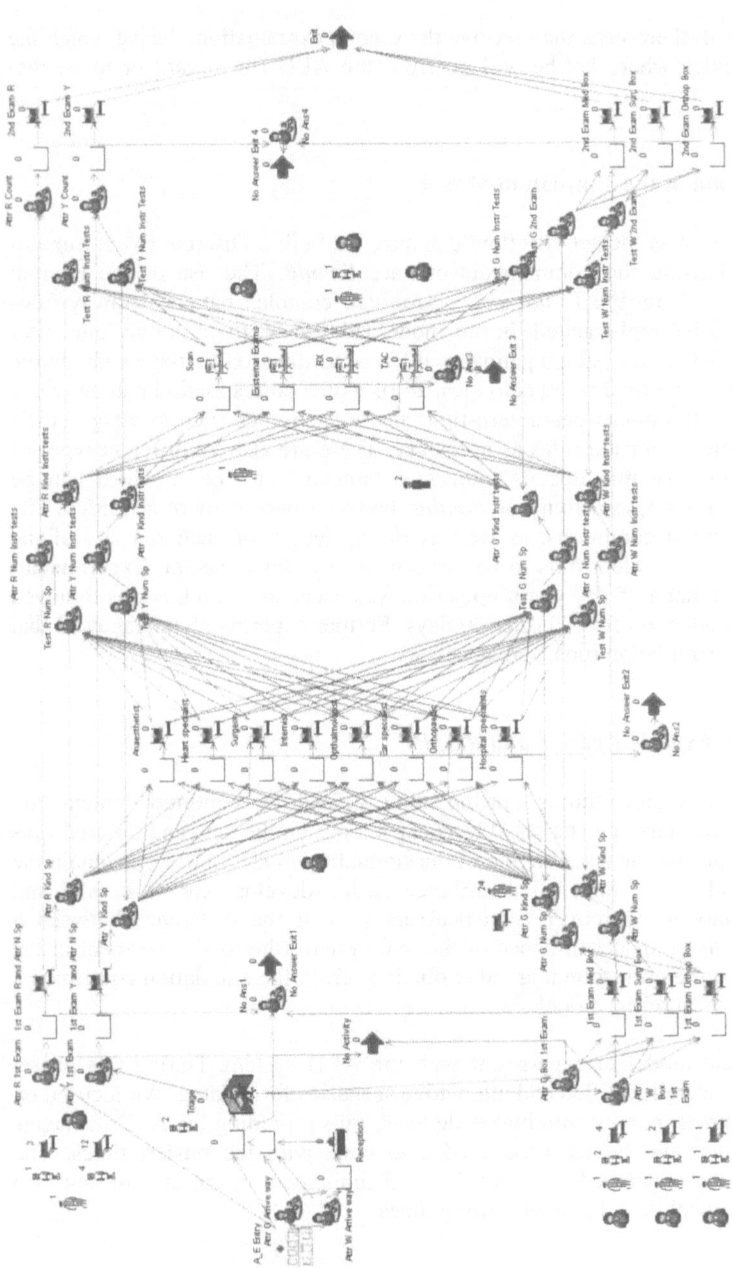

Figure 5: Simulation Model (*Simul8* screen-shot showing patient pathways through the AED).

Table 1 shows some performance measures of the utilization of the different categories of doctors working in the AED. The percentages of the utilization time for the radiologist dedicated to CAT (*Computerized Axial Tomography*) and the orthopaedist are the highest.

MEDICAL RESOURCES	**Utilization Time %**	**Average Use**	**Maximum Use**	**Maximum Availability**
Radiologists CAT_Scan	**69.48**	1.04	2	2
Orthopaedist	**68.73**	0.69	1	1
Radiologists X-ray	51.82	0.52	1	1
Surgeons	50.92	0.51	1	1
Emerg. Doctors (green/white)	46.68	0.81	3	3
Ear specialist	45.78	0.46	1	1
Opthalmologist	41.29	0.41	1	1
Emerg. Doctors (red/yellow)	38.00	0.76	3	3
Heart specialist	23.02	0.23	1	1
Anaesthetist	4.87	0.49	1	1
Hospital Specialists	3.40	0.81	6	24

Table 1: Performance measures for resource utilization.

Table 2 presents the results for the queues for the different activities, showing critical situations in the waiting time (in minutes) for the first orthopaedic examination of patients with green or white codes, and for CAT scans and X-ray.

These observations led to the formulation, simulation and analysis of three alternative scenarios. In the first scenario, patients with a white code were removed from the AED (to be seen elsewhere as they are the most minor category). In the second scenario, an extra CAT and radiologist were added, and in the third scenario an orthopaedist and an additional examination cubicle were added.

	Average Queuing Time	**Maximum Queuing Time**	**Average Queue Size**	**Maximum Queue Size**
Queue for 1st exam Orthop	**138.62**	240.00	4.78	21.71
Queue for CAT	**92.74**	180.00	3.35	16.80
Queue for Orthopaedist	**22.43**	180.00	0.77	9.21
Queue for X-ray	**16.11**	180.00	2.97	25.01
Queue for Ear specialist	11.74	174.57	0.33	7.18
Queue for Opthalmologist	9.22	152.07	0.23	6.02
Queue for Surgery	7.62	189.89	0.24	9.34
Queue for Surgeon	7.13	131.69	0.12	4.47
Queue for Scan	5.48	160.29	0.16	5.58
Queue for Heart specialist	4.13	95.42	0.06	3.89
Reception Queue	3.03	36.26	0.37	10.14
Queue for 2nd Exam Surgery	3.00	120.12	0.09	6.83
Queue for 2nd Exam (red)	1.55	20.98	0.00	1.06
Triage Queue	0.93	72.46	0.15	7.33
Queue for 2nd Exam Orthop	0.80	28.45	0.02	2.91
Queue for Anaesthetist	0.68	21.16	0.00	1.29
Queue for Internist	0.49	8.95	0.00	1.03
Queue for 1st Exam Med	0.49	59.90	0.03	5.85
Queue for Hospital specialists	0.28	52.64	0.01	4.11
Queue for 2nd Exam Med	0.02	8.41	0.00	2.23
Queue for 2nd Exam (yellow)	0.00	2.46	0.00	1.11
Queue for External Exams	0.00	0.00	0.00	1.00

Table 2: Queue statistics

Figure 6 shows the resource utilization for each of the three scenarios compared with the current situation (the baseline). In the first scenario, the utilization of all resources decreases slightly. In the second scenario the utilization of the CAT radiologist decreases further, and in the third scenario there is an even greater decrease in the orthopaedist utilization, so that it approaches the same value as the other resources.

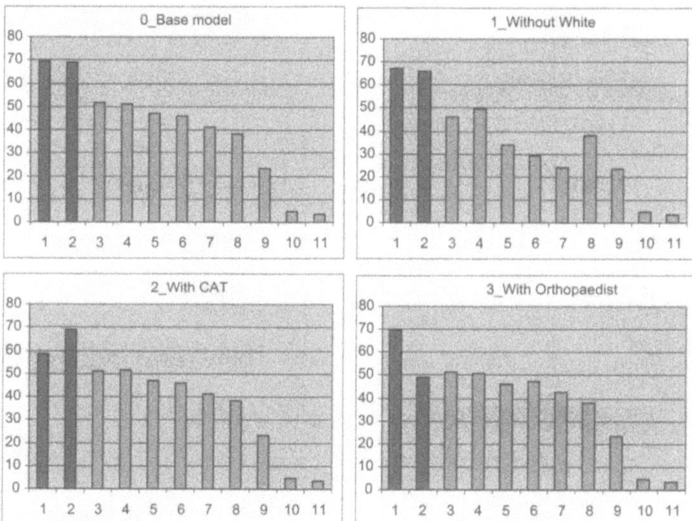

1 = Radiologists CAT_Scan, 2 = Orthopaedist, 3 = Radiologists X-ray, 4 = Surgeons, 5 = Emerg Doctors (green/white), 6 = Ear specialist, 7 = Ophthalmologist, 8 = Emerg Doctors (red/yellow), 9 = Heart specialist, 10 = Anaesthetist,

Figure 6: Comparison of the resource utilization in the baseline and 3 scenarios.

Figure 7 shows the waiting times in each of the three scenarios, compared to the current situation (the baseline). In the first scenario there is a slight decrease in the average waiting time in all the queues. In the second scenario, waiting times change only for the CAT queue. In the third scenario there is a significant reduction in the waiting times for the first orthopaedic examination for green and white code patients. Here the reduction is from 138 minutes to 19 minutes, while the waiting times in all the other queues do not change.

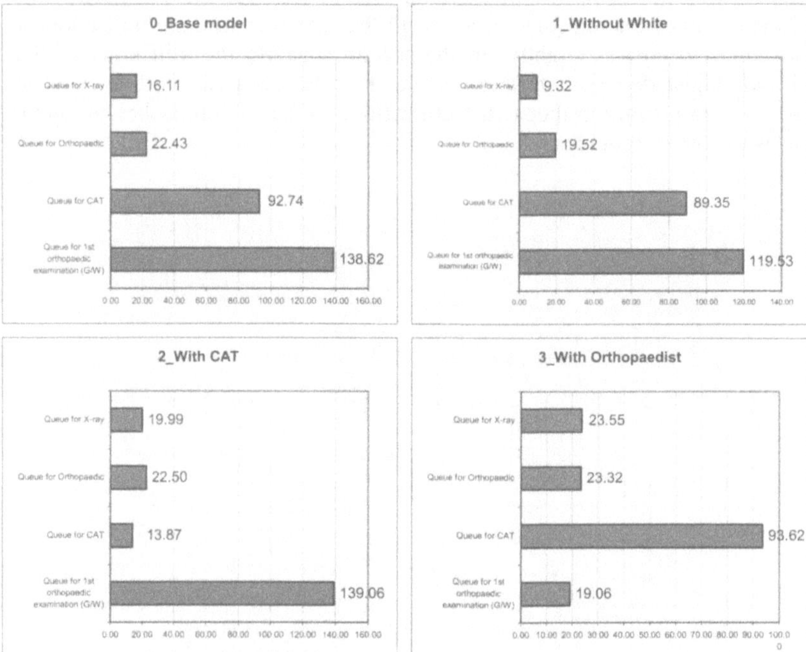

Figure 7: Comparison of the average waiting time (in minutes).

Conclusions

This work enabled us to build a simulation model that describes the working of a complex Emergency Department, one of the largest in Rome. By analysing the AED processes we were able to understand the system as a whole, and to build a logical-mathematical-statistical model to represent its objects, their attributes and their functional relationships. A discrete-event simulation model has been implemented using the software *Simul8*, and the simulation results have allowed us to identify the critical parts of the AED. It has been possible to formulate and evaluate alternative scenarios by changing parameter values, thus improving the performance of the system. Access to more detailed data could potentially improve the accuracy of the results, but the model validated well against real data and was a useful tool to inform the AED managers how to eliminate significant bottlenecks. Further work will use this baseline tool to simulate the response of the system to high emergency situations due to exceptional events.

References

Bollettino Ufficiale della Regione Lazio (1994), Deliberazione del Consiglio Regionale 11 maggio 1994, n. 1004, "Sistema di emergenza sanitaria Lazio Soccorso 118", in: Bollettino Ufficiale della Regione Lazio n. 21 del 30/07/1994 - Parte prima.

Bollettino Ufficiale della Regione Lazio (1997), Deliberazione della Giunta Regionale 8 luglio 1997, n. 4238, "Legge 18 luglio 1996, n. 382 - Piano di ristrutturazione della rete ospedaliera regionale nel triennio 1997–1999", in: Bollettino Ufficiale della Regione Lazio n. 25 del 10/09/1997 – Parte prima.

Bollettino Ufficiale della Regione Lazio (2002), Piano Sanitario Regionale per il triennio 2002-2004, in: Supplemento Ordinario n. 7 del Bollettino Ufficiale della Regione Lazio n. 24 del 30/08/2002.

Bowers J., Mould G., Whitehead C. (2000) "An Examination of the Organisational Impact of Re-designing Clinical Services – Orthopaedic Simulation", in: J. Riley (ed.), Planning For the Future Health, Service Quality and Emergency Accessibility, Glasgow Caledonian University: 3–17.

Brailsford, S. C., Lattimer, V. A., Tarnaras, P. and Turnbull, J. A. (2004). Emergency and On-Demand Health Care: Modelling a Large Complex System. Journal of the Operational Research Society, Vol 55, 34–42.

De Angelis V., Felici G. and Storchi G. (2004), "Planning the emergency ambulance service in the city of Rome by a mixed integer linear programming model", in: M. J. Ferreira De Oliveira (ed.), Accessibility and Quality of Health Services, Frankfurt Am Main: Peter Lang (Germany), pp. 35–45.

De Angelis V., Felici G., Impelluso P. (2003), "Integrating simulation and optimisation in health care center management", European Journal of Operational Research, 150 (1): 101–114.

Facchin P., Romanin-Jacur G (2003), "Emergency service: a generalised flexible simulation model", lavoro presentato al convegno AIRO.

Gazzetta Ufficiale (1992), Decreto del Presidente della Repubblica 27 marzo 1992, "Atto di indirizzo e coordinamento alle regioni per la determinazione dei livelli di assistenza sanitaria di emergenza", in: Gazzetta Ufficiale n. 76 del 31/03/1992.

Gazzetta Ufficiale (1996) "Atto di intesa tra Stato e regioni di approvazione delle linee guida sul sistema di emergenza sanitaria in applicazione del Decreto del Presidente della Repubblica 27 marzo 1992", in: Gazzetta Ufficiale n. 114 del 17/05/1996.

Gazzetta Ufficiale (2001) "Linee Guida sul 'Triage intraospedaliero' per gli utenti che accedono in Pronto Soccorso, emanate dal Ministaro della Salute con i Presidenti delle Regioni", in: Gazzetta Ufficiale n. 285 del 7/12/2001.

Harper PR (2002), "A Framework for Operational Modelling of Hospital Resources". Health Care Management Science. 5 (3): 165–173.

Harper PR and Shahani AK (2003), "A Decision Support System for the Care of HIV and AIDS Patients in India". European Journal of Operational Research. 147 (1): 187–197.

Lagergren M., "A simulation model concerning future needs of long-term care of elderly persons in Sweden", in: M. J. Ferreira De Oliveira (ed.), Accessibility and Quality of Health Services, Frankfurt Am Main: Peter Lang (Germany), 21–31.

Law AM. and Kelton WD. (1982), Simulation Modeling and Analyses, McGraw-Hill, New York..

Lubicz M. and Mielczarek B. (1987), "Simulation modelling of emergency medical services", European Journal of Operational Research, 29: 178–185.

Romanin-Jacur G. and Facchin P (1987), "Optimal planning of a paediatric semi-intensive care unit via simulation", European Journal of Operational Research, 29:192–198.

Vissers J.M.H., Croonen G., Siersma V., Tiemessen H. (1999) "Simulation of the cardiology patients flow in a hospital setting", in: V. De Angelis, N. Ricciardi and G. Storchi (eds.), Monitoring, Evaluating, Planning Health Services, World Scientific, 162–173.

WAITING TIME AND REDUCING THE DELAY IN HOSPITAL ELECTIVE PROCEDURES

SAEEDEH KETABI
Department of Management, University of Isfahan, Iran

Abstract

One of the major issues in health care is the waiting time to access services such as hospital care. In this paper the whole process of admission–treatment–discharge is investigated in a large teaching hospital in Iran. The process may be formulated as a network of queues. The productivity factors are calculated by simulation, and then the parameters that influence the delays in the process are studied. It seems that the bottleneck in the process lies in the discharge process, i.e. from the discharge consent until the patient actually leaves the hospital. Finally, several managerial decisions which reduce the delay are discussed.

Keywords: admission-treatment-discharge process; length of stay; queueing networks; simulation.

Introduction

Health care management is a very complicated and demanding business. Limitations in health care funding require physicians and hospitals to find effective ways to utilize resources. Although the health care industry faces many of the issues which confront other industries, it is a unique business in many respects.. It has multiple decision-makers with conflicting goals and objectives (Li and Benton, 1996). The health care industry has objectives, costs, customers and patterns of behaviour that differ from other industries. In health care management the main objectives are to increase quality of and access to health care services and, at the same time, to control spiralling medical costs. A fundamental aspect of health care management is the effective allocation of resources (Kusters and Groot, 1996). Operational Research methods can be used to improve the performance and efficient use of existing resources; to minimize costs and maximize quality, or to introduce changes which combine both.. One of the major issues in health care is the waiting times for patients to access services such as hospital care.

Hospitals are the most important unit in the health care industry, as they are a frontline service for the public. Patient satisfaction is one of the most important factors in evaluating health care services. Delay in accessing health care is not only undesirable for patients from a financial and psychological point of view, but is also undesirable for health managers, due to the high cost of intermediate health services and/or additional costs incurred due to a worsening condition. Hospitals are faced with limited resources, which sometimes means insufficient beds to facilitate the care of admitted patients.

In this research, hospital processes are studied to provide a decision-based model for the admission-treatment-discharge procedure. It will then be possible to investigate alternative ways to reduce the waiting times of elective patients, who are to be admitted to the hospital. The proposed model is built on a foundation of two months daily data collection from two wards, Gastro-enterology and Paediatric surgery, in Azzahra hospital, a large, specialised teaching hospital in Isfahan, Iran. In Iran, there are differences between teaching and non-teaching hospitals, mainly in the care process and the scope of the managerial decisions. Each ward has 28 beds for elective and emergency patients. Although the overall bed-occupancy ratio is not very high, there are days when patients have to wait several hours or even days for a vacant bed. It is important to investigate the causes of the delays in admission and also the discharge processes, since they affect the patient's length of stay at the hospital, *i.e.* deal with the hospital resource allocation issues and improve efficiency. The whole process of admission-treatment-discharge is investigated, although the focus will be on the discharge process.

Since health care services have a stochastic nature, the processes can be thought of as a network of queues. Two quantitative models are suggested: a queueing theory model and a discrete event simulation. Unfortunately, the data is not statistically appropriate for applying analytical queueing methods. Therefore the simulation model will be established on the queueing model to evaluate the current performance and to test the impact of changes. Simulation involves the generation of an artificial copy of the system through an abstract model of the real world situation. The analysis allows us to draw conclusions about the operating conditions in the real world system. The simulation models are more suitable for complicated systems in which any changes for improvement are costly with non-deterministic results

Admission–Treatment–Discharge Process

Azzahra hospital in Isfahan, one of the largest health care centres in the Middle East, opened in 1993. It consists of 35 wards over 6 floors, housing 750 active beds. Two wards, Gastroenterology and Paediatric surgery, have been selected for investigation. Both wards face admission procedure delays, although there are major differences in their performance. The admission and discharge (Income office) procedures of two departments are studied, particularly as they increase the delay in gaining access to the health services. Figure 1 summarizes the admission–treatment–discharge process. The procedure is more complex than shown in the work flow diagram, i.e. there is the flow of patients to and from other wards in the admission and discharge sub-processes. Also, when interviewing the staff, variations were revealed, for instance, procedures are not followed in exactly the same order by patients or staff. Therefore, it seems that likely that understanding and redesigning the process will result in increased productivity.

The rate at which patients arrive to be admitted to wards varies according to the season, day and time of day. Patients are admitted in accordance with the physician's referral. The initial patient care or pre-assessment is automatically undertaken by the nurses in the Paediatric surgical ward, while patients in the Gastroenterology ward must wait for the doctor to visit.. This is a significant difference between the two wards. Another difference is in the visiting schedule for doctors and their interns. The Paediatric surgical ward has an early morning schedule for the doctor's visit, which ensures the discharge process starts earlier in the day. On the other hand, the interns have no schedule, which can cause a delay in writing up the abstract medical report, which is a part of the discharge process. The Gastroenterology ward has no specific schedule for doctors or interns to visit, sometimes resulting in a treated patient waiting hours or even days for a doctor to issue a consent for discharge.

A queueing network model

Models are used as a powerful tool in health service management and research. Managers, who are striving to improve performance, can use information from modelling to help them decide on how existing resources could be used in a more efficient way. Modelling aims to increase the understanding of how systems operate and make it possible to predict the system response to a range of different changes, especially changes which may initially appear either impossible or too costly to be implemented.

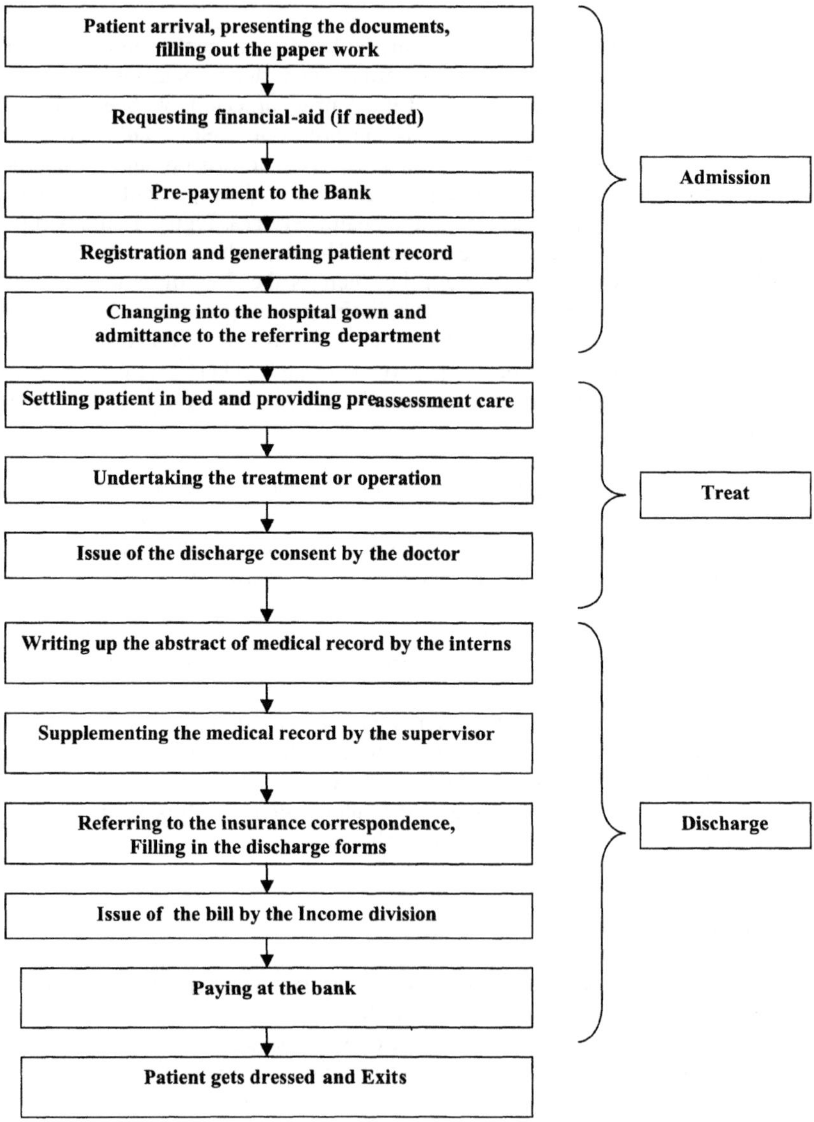

Figure 1: Work flow in the hospital elective process

It is shown that the queueing model is a potent instrument that can be used by the health care planner overseeing the waiting system, in order to maximize patient welfare whilst minimizing healthcare costs. It should be noted that waiting systems for medical services have distinct features that differentiate them from the queueing systems commonly studied in Operational Research, i.e. the variability in the quality of service, variability in the duration of service, and variations in patient needs.

The admission-treatment-discharge process may be viewed as a network of queues as shown in Figure 2:

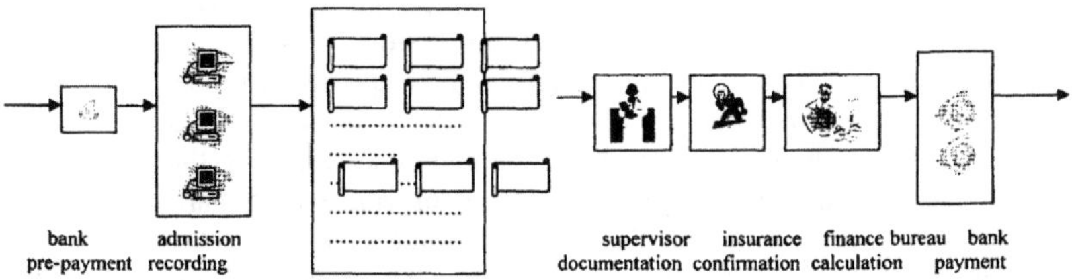

Figure 2: The queueing scheme for admission-treatment-discharge process

The choice of modelling approach is mainly determined by the nature of the problem being studied. One of the main difficulties in applying an analytical method is the lack of the required data. In this research, the data was collected retrospectively, and thus the corresponding data for the related activities were not accurate enough to ensure a reliable queueing model. The reasons for this are firstly some data is missing, secondly the fact that statistical assumptions, such as independence, did not hold, and thirdly the variation in the stages which patients followed in the process. Many researchers in the health management area agree that simulation models permit more flexibility and enforce less unwanted simplification than the analytic models (Swisher *et al.*, 2001). A disadvantage with stochastic simulation is that it requires statistical analysis of the outcome and often leads to imprecise conclusions. On the other hand, it must be noted that models in the health area need to be simple and transparent in order to be understood by healthcare managers who may be unfamiliar with the concepts.

The Quantitative Approach

The study of Azzahra Hospital waiting time consisted of three phases: data collection, identification of impact factors and the suggestion and testing of the Operational scenario.

Data collection

A major challenge in building quantitative models lies in the initial data collection and analysis. In this research, the data were extracted from both the computer database and written medical records. The basic patient information such as, age, gender, insurance company, physician, arrival time at the admission office and subsequent ward admission, was collected from the computer database. The main data required for a queueing model, such as the time that the patient was first seen by a physician or pre-assessed by a nurse and the time that the doctor approved the patient discharge (or the time of the patient's death) had to be extracted from the relevant paper (hard) record. The length of stay for each patient has been calculated by subtracting the time of admission from the time of leaving hospital. The discharge delay has been defined as the time difference between the granting of permission for the patient to be discharged and the time the patient left the hospital.

In this research the statistics of patient demand (arrivals), length of stay (service time) for different patient groups and time spent in the admission and discharge sub-process have been established (Table 1, 2 and 3). Timing of the admission and discharge sub-process includes the corresponding waiting times, so that the related statistical quantities in the last column of Table 3 can be a fine approximation. The model is thus able to take into consideration the time consumed in the Admission and Income Office by patients from the other wards.

Patient group	Arrival rate	Average length of stay	Standard deviation of los	Average los without the discharge time	Standard deviation of los without the discharge delay
1	1.84	0.72	0.45	0.69	0.35
2	1.38	2.27	0.45	1.71	0.59
3	1	5.02	1.12	4.95	1.18
4	0.31	13.45	9.21	9.54	2.38

Table 1: Statistical quantities (per day) for Pediatric surgery Ward

The two wards have then been statistically compared, with both descriptive and perceptive statistics being obtained using SPSS. There are differences in administrative procedures in the two wards; mainly in the planning of the doctors' visits. It has been shown that the delay in discharge is not the same in both wards, indicating that the planning of the doctors' visits affects the discharge delay process. The amount of time that an elective patient may have to spend in the hospital prior to release is stochastic. Planned patient admission profiles are highly correlated with the time and day of the week, clearly because of the doctors' schedules. It was confirmed that the arrival of patients is Poisson distributed, with the rate depending on the day of week. A chi-square test was carried out to check the significance of the relationship between the patient's attributes and length of stay (LOS). Different patient types have different LOS distribution profiles. It is noted that the patient groupings were derived within SPSS and were confirmed by the Kruskal-Wallis test. Some factors that define the length of stay of the patient relate to information peculiar to the individual patient, such as age, physician, arrival time and the diagnosis. The information may be known prior to admission for a majority of patients, which may provide useful information to the hospital in terms of matching resources to planned admissions.

Patient group	Arrival rate	Average length of stay	Standard deviation of los	Average los without the discharge time	Standard deviation of los without the discharge delay
1	0.66	0.9	0.3	0.695	0.28
2	0.75	2.63	0.49	2.4	0.54
3	0.97	5.17	1.11	4.87	1.17
4	0.66	13.78	7.31	13.48	7.62

Table 2: Statistical quantities (per day) for Gastroenterology Ward

	Paediatric surgery	Gastro-enterology	General average
Number of referral patients	280	190	
Average of admission time	2:12	1:08	0.48
Standard deviation of admission time	2:14	1:23	0.48
Average of discharge delay	6:09	4:04	1.2
Standard deviation of discharge delay	7:00	2:42	0.96

Table 3: Statistical quantities for admission and discharge departments

Evaluating length of stay information is a challenging task, but is essential for the operational success of a hospital. An estimate of occupancy could help in scheduling future elective surgeries, some of which are currently postponed due to lack of bed availability. Predicting LOS is difficult and is often only done retrospectively (Swisher *et al.*, 2001). A patient's LOS is correlated with the severity of their illness.. Prediction of the degree of care required by a patient naturally separates the patient population into several distinct categories. Marshal *et al.* (2002) estimates LOS by categorizing the patients to facilitate resources. Xiao *et al.* (1999) propose a mixture distribution for LOS. Ridge *et al.* (1997) state that the length of stay of emergency patients and planned patients are either negative exponential or Weibull distributed, but this was not verified for our data. One research limitation was the low reliability of the retrospective data. Therefore, for those statistical variables for which we could not fit an Exponential, Gamma or Log-Normal distribution function, we used a general distribution function with the experimental mean and variance.

Impact factors identification

Observing the related procedures in Azzahra Hospital and interviewing the staff and patients, reveal that there are several factors influencing the delay in accessing the services. These factors are as follows:

- Lack of co-ordination between the admission department and the wards and no access to the updated databases
- Bank closing (from time to time)
- Shortage of hospital gowns and the pre-assessment nurse (from time to time)
- Shortage of bed (from time to time)
- Lack of information for the patients about the process and the required documents
- Incomplete medical records
- Delay in writing the medical abstract

Operational suggestions

The impact factors which are responsible for the delay to patients in getting access to a hospital bed are found in all parts of the admission-treatment-discharge process. Therefore, changes may be formulated for the Admission Department, wards and Income Office to obviate these factors. The following strategies are simple and easily implemented.

Admission Department:

- Precise and timely co-ordination between the different departments.
- Undertaking the diagnostic and para-clinical stages prior to admission.
- Plan a standard referral procedure for doctors, when referring patients for hospital treatment, i.e. planning a queuing system for reservations.
- Regulate the number of staff in the Admissions Office and/or bank according to the rate of patient arrivals.
- Provide clear signage to direct patients to relevant areas.

Wards:

- Research the average length of patient stay.
- Predict the demand, i.e. number of patients arriving and adjust the number of beds accordingly.
- Arrange for nurses to undertake initial routine care or examination before the doctor visit.
- Keep thorough, ongoing and up to date patient medical records
- Review the qualifications and managerial standards needed for the ward managers and supervisors
- Plan and schedule the doctors' visits to patients.
- Provide a master schedule for doctors' visits to patients, to ensure that interns are available to write up medical records.
- Improve communication and co-ordination between departments.

Income Office:

- Inform patients about the discharge procedure and the required documents.
- Undertake discharge steps such as insurance confirmation, the calculation of surgery expenses, and blood bank costs prior to actual discharge consent being received.
- Calculate bed charges at an hourly rate rather than a daily rate to encourage patients to leave quickly after their discharge is agreed.
- Provide unstaffed waiting areas for patients while they are waiting for the discharge process to be completed.
- Provide a schedule for each ward to refer patients to the Income Department.

A common practice in hospitals is for the arrival of elective patients to be scheduled according to their position on the waiting list.. Scheduling patients in a different way could result in a better balance of arrivals allowing hospitals to allocate their resources more effectively. It is clear that other factors also affect productivity and waiting times, but their effect has not yet been quantified, apart from two, adjusting the number of beds and providing unstaffed waiting rooms for pending discharged patients. This will be discussed more in the next section.

Results

In this paper the admission–treatment–discharge process has been described and analysed in Azzahra Hospital in Iran, to determine the impact factors on the delay that patients are face to access elective treatment. Some operational scenarios have been suggested based on observational studies. Two simulation models have been established based on the queueing model for the wards. WinQsb (Chang and Desai, 1996) is used to simulate events with probability data. The simulation horizon is 300 days, data has been collected from the 200th day and the results have been calculated as the average of 30 runs of the model. The validity of the model has been verified by statistically comparing the average length and waiting time of the queues generated by the model with the observed values. The simulation results are summarized in Table 4 and 5. Table 4 summarizes the patient groups' statistics. Table 5 shows the effect of varying the number of beds, and the effect of eliminating the discharge delay by transferring the treated patients to unstaffed rooms ("without discharge delay").

Patient group		**1**	**2**	**3**	**4**
Paediatric surgery	No. of arrivals	185	149	116	25
	Ave. process time	1.22	2.41	6.35	11.87
Gastroenterology	No. of arrivals	60	67	112	61
	Ave. process time	1.32	3.31	6.38	21.45

Table 4: Simulation results I

	ρ	L	L_q	W_q
Paediatric surgery ward with 28 beds	66.2%	20.78	0.29	0.06
Paediatric surgery ward with 29 beds	61.8%	20.63	0.24	0.05
Paediatric surgery without discharge delay	52.8%	16.02	0.20	0.04
Gastroenterology ward with 28 beds	77.2%	24.71	1.67	0.56
Gastroenterology ward with 29 beds	70.8%	20.97	0.07	0.05
Gastroenterology without discharge delay	78.6%	24.07	1.12	0.36

ρ : System utilization factor, L : Average number of customers in the system
L_q: Average number of customers in the queue, W_q: Average time customer spends in the queue

Table 5: Simulation results II

The results show a very low bed utilization factor and short waiting time, contrary to the observation made that there are days when patients have to wait several hours or days for an empty bed This confirms that the current bed number could satisfy the patient demand, but that there is mismanagement in the hospital procedures for patient arrival and discharge procedures.

As the *Data Collection* Section and Table 3 show, the discharge delay is significantly different between the two wards. Although the Paediatric surgical ward achieves a better standard of efficiency due to the implementation of some of the operational suggestions proposed in the *Operational Suggestions* Section, statistical quantities show the discharge delay for the Gastroenterology ward is lower. In the Gastroenterology ward, the long delay in receiving discharge permission from the doctors does not affect the discharge process, which is not commenced until after discharge permission has been issued. In the Paediatric Surgical ward, despite the planned doctor visiting schedule, the interns' delay in writing up the medical abstract makes the discharge delay longer. Therefore, removing the discharge delay decreases the beds' utilization factor. In the Gastroenterology ward, the discharge delay, as defined in this paper, does not affect the utilization factor; it is important not to confuse the delay in issuing the discharge permission and the actual length of stay.

The procedure is more complex than has been showed in the work flow diagram and simulated in this section; there is the flow of patients from other wards in the admission and discharge sub-processes. Also the related activities are not followed in exactly the same order by patients or staff. To ensure a more reliable model, the data needs to be collected by actually tracking the patients, rather than obtaining data retrospectively.

Conclusion

This paper presents a queueing simulation model and demonstrates the power of using operational research methods in evaluating hospital performance when trying to reduce patient delays and improve productivity and efficiency, in the elective patient procedures. Also, the impact factors that influence delay in the process have been studied. The collected data was not suitable to apply the queueing model. The analytical queuing model is best suited to a health care unit which provides very specific and definite services, such as medical diagnostic centres, laboratories or blood banks. The initial requirement to produce an optimization model in hospitals is a coordinated information system which is continuously updated.

Because the demand made on a hospital's services are random and because the services times provided are random, it is not possible to avoid all delay, but it is the task of the hospital management and administration to reduce it to a minimum. Hospitals are faced with multiple-objective decision making problems and need to employ better tools, together with a structured system of forecasting-planning-scheduling. It becomes clear that as Vissers (1998) a state, knowing the patients' arrival distribution provides flexibility in allocating facilities (such as beds) to different departments. For more precise planning, it is also necessary to have a more accurate database for LOS rather than the general information which is considered in this research.

The outcome of applying the following managerial decisions in different departments on reducing delay time has been explored:

1. Availability of a complete, accurate and up to date hospital database.
2. A master schedule for doctors to visit patients and for interns to write up the medical records.
3. Initial care managed by nurses.
4. A continuous updating of medical records,
5. Providing unstaffed waiting rooms for patients prior to discharge.
6. Predicting the future hospital resource need.
7. Predicting the length of stay for each category of patients.
8. Scheduling elective patients to even out workloads and maximize use of resources.

As can be seen from the simulation result, eliminating discharge delays decrease the waiting times. The discharge process, i.e. from when the doctor issues the discharge permission until the patient leaves the hospital, is a bottleneck and raises the bed occupancy rate as treated patients cannot leave the hospital until end of discharge process, which in turn increases the wait for the patient, either the arrival patient wait for a free bed or the treated patient for discharge-letter) Indeed, complex and unnecessary activities, lack of relevant and consistent information, shortage of clerical staff, lack of a sequence of activities in discharge process causes major delay, conflict and dissatisfaction of the whole process. Therefore, it is suggested that a research study is carried out on the discharge procedure and a detailed model produced for the discharge process only, which will be complicated since all wards will be involved in the data collection.

OR can currently do a great deal, but we are a long way from what could and will be done in the future. A requirement to achieve this is lose collaboration between the clinicians and OR researchers. The health care units in Iran are subject to many informal rules and activities, which are often random. As procedures in Azzahra Hospital have been studied, so could many other health care processes be studied using simulation. Even the study of the work flow diagrams of the processes and the application of business process re-engineering will almost certainly bring about a big improvement in waiting times.

References

Kusters, D. and Groot, P., "Modelling resource availability in general hospitals, design and implementation of a decision support model", European *Journal of Operations Research* (88), pp 428–445, 1996.

Li, L.X. and Benton, W.C., "Performance measurement criteria in health care organizations: review and future research directions", *European Journal of Operations Research* (93), pp 449–468, 1996.

Marshal, A.H. and McClean, S.I. and Shapcott, C.M. and Millard, P.H., "Modelling patient duration of stay to facilitate resource management of geriatric hospitals", Health Care Management Sciences (5), pp 313–319, 2002.

Ridge J.C., Jones S.K., Nielsen M.S. and Shahani A.K., "Capacity planning for intensive care units", European Journal of Operational Research (105), pp 346–355, 1997.

Swisher, J.R. and Jacobson, S. H. and Jun, B. and Balci. O., "Modeling and analyzing a physician environment using discrete-event (visual) simulation", Computers and Operations Research (28), pp 105–125, 2001.

Vissers, J.,"Patient flow-based allocation of inpatient resources: A case study", *European Journal of Operations Research* (105), pp 356–370, 1998.

Xiao, J and Lee, A.H. and Vemuri, S.R., "Mixture distribution analysis of length of hospital stay for efficient funding", Socio-economic planning Sciences (33), pp 39–59, 1999.

Yih-long Chang and Kiran Desai, WinQsb version 2.0 – Decision software for management science and operations management, Wiley, 1996.

OPTIMISING THE USE OF RESOURCES

DECISION SUPPORT FOR EFFICIENT AMBULANCE LOGISTICS

TOBIAS ANDERSSON[1], SVERKER PETERSSON[2]
and PETER VÄRBRAND[1]
[1]*Division of Communications and Transport Systems, Department of Science and Technology, Linköping University, Sweden*
[2]*SOS Alarm AB, Sweden*

Abstract

Ambulance logistics comprises everything that has to do with managing an efficient ambulance health care, which includes both emergency medical services and patient transportations. In this work, ambulance logistics is defined and discussed using the Swedish public service enterprise SOS Alarm as a basis. A number of ways of improving the ambulance logistic service provided by SOS Alarm are identified, and nine areas of improvement are defined. Two of the nine areas, "Visualisation of the preparedness" and "Decision support for efficient ambulance logistics" are described in more detail. In both of these, the main improvement lays in the development of decision support tools for the ambulance dispatchers. The tools that are developed are a preparedness calculator, an ambulance dispatch suggestion tool, a relocation tool and a simulation tool.

Keywords: health; logistics; optimization; simulation; decision support systems; heuristics.

Introduction

An emergency medical services system commonly includes ambulances, but this does not necessarily make ambulance health care the same thing as emergency health care. Ambulance health care does include urgent out-of-hospital medical treatment, as well as possible transportation of the patient. However, it may also include the transportation of patients not in need of advanced or urgent medical care. Simply put, ambulance health care encompasses all services performed with an ambulance, and ambulance logistics is the process of managing these services. In Sweden, the foremost provider of an ambulance logistics service is the company SOS Alarm AB.

In this paper, ambulance logistics will be defined and discussed using SOS Alarm and the services the company supplies, as a starting point. Furthermore, a number of areas where the ambulance logistics service can be improved are identified and presented. Finally, work that has been performed in two of these areas is presented. This concerns the development and testing of a number of decision support tools for the ambulance dispatchers.

The paper is intentionally low on technical details, and its main purpose is to be an easily read overview of a project within ambulance logistics, as well as an introduction to the area.

Ambulance Logistics

The Swedish National Board of Health and Welfare defines *Ambulance* as (Socialstyrelsen, 2001, translated from Swedish):

> *"Means of conveying a patient, which can be a car, a boat, a helicopter or another kind of specially equipped vehicle."*

The Council of Supply Chain Management Professionals defines *Logistics Management* as [http://www.cscmp.org/Website/AboutCSCMP/Definitions/Definitions.asp, accessed 2005-09-29]:

> *"Logistics Management is that part of Supply Chain Management that plans, implements, and controls the efficient, effective forward and reverse flow and storage of goods, services and related information between the point of origin and the point of consumption in order to meet the customers' requirements."*

Using these two definitions as a base, a definition for *Ambulance Logistics* is suggested below. The definition of ambulance makes it clear that it is a vehicle that can be used to transport a passenger, i.e. the patient. The patient is the customer, and a general objective in logistics management is to satisfy the customer. Thus, "to meet the customers' requirements" will be to take care of the patient in a satisfactory manner. Intuitively, "the point of consumption" from the patient's point of view means the actual transportation. More accurately it could be when the ambulance personnel have reached the patient and started the medical treatment. This however implies that driving the patient to his or her destination is not a part of the logistic process, which is clearly not true; the ambulance logistics process stretches at least to the point when the ambulance personnel have delivered the patient and are once again available for new assignments. Therefore, it is more appropriate to define the point of consumption as the time when the patient has no further need of the ambulance. That is for example when the ambulance has left the patient at the correct destination, or it has been determined that the patient did not need an ambulance in the first place.

"The point of origin" is when the request for an ambulance is received, and the "flow and storage of goods, services and related information" that takes place between the point of origin and point of consumption can in ambulance services be translated to:

- *flow of goods:* The main goods that are handled are the patients, though it is possible, and perhaps less controversial, to view the transportation as a service. The goods can also be resources such as medical personnel and equipment.
- *flow of services:* Besides the transportation being a service, the medical treatment of the patients is a major part of ambulance logistics.
- *storage:* The main storage, or buffer, consists of available units, i.e. ambulances that are waiting to be dispatched. It is also necessary to store consumable medical equipment, such as oxygen.
- *information:* It is crucial that there exists an efficient information flow between ambulance units and dispatchers. It is also important that it is possible to derive information from the patient.

Using the definition of logistics management above as a starting point, it is possible to define *Ambulance Logistics* as:

> *Ambulance Logistics is the planning, implementation, and control of resources and information used to facilitate an efficient way of serving a person in need of out-of-hospital medical care including possible transportation.*

Thus, according to the definition above, ambulance logistics covers more or less all aspects of maintaining an efficient and effective ambulance health care.

In Sweden, the foremost supplier of ambulance logistics is SOS Alarm. SOS Alarm is a Swedish public service enterprise, owned by the government (50%), the county councils (25%) and the municipalities (25%). The company is responsible for receiving all calls to the national emergency telephone number, 112, which is used when life or property is in danger. Dialling this number it is possible to get in touch with ambulance services, fire services, police, medical expertise, sea or air rescue, priest on call and a few other services. Of the rescue services, SOS Alarm is involved in the planning and control of the majority of the ambulances and fire service resources in Sweden. Commercial alarm services and after hours telephone services are also offered.

The services are run from an SOS centre. There are currently 20 centres in Sweden, roughly one for each county. The ambulance logistics service at SOS Alarm is schematically described in Figure 1. It can be divided into four functions; the Order handling function, the Ambulance planning function, the Ambulance control function and the SOS Waiting room function.

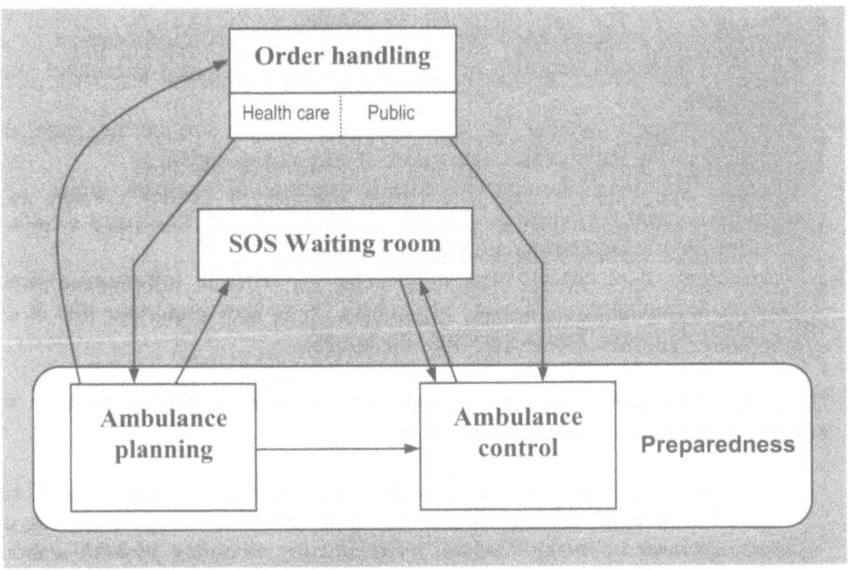

Figure 1: A schematic picture of the ambulance logistics service provided by SOS Alarm

It should be noted that an SOS centre does not have to be organised as indicated by Figure 1; the same person may be responsible for all functions in Figure 1, something that is common in smaller centres. On the other hand, in larger centres there may be several people working in one of the functions, for example the Ambulance control function, where different dispatchers may handle calls of different priority or may be responsible for different parts of the county.

The whole process, from someone requesting an ambulance, until the ambulance has treated, transported and left the patient, is referred to as an *ambulance call.* The call is *received* in the *Order handling function* when an SOS operator answers the phone. The call may be urgent, which is most often the case when it is made through the emergency number, typically by the public. Non-urgent ambulance calls can be received through the emergency number, but also through a special ambulance request number that is commonly used by the health carers. Regardless of how or from where the call is received, it is the task of an SOS operator to judge the urgency of the call. This is referred to as *prioritising* the call. The operators have a medical index manual (Svenskt index för akutmedicinsk larmmottagning, 2001), consisting of a number of standardised questions and criteria that help them decide which priority an incoming call should have. The calls are prioritised according to three degrees:

- Prio 1: Urgent, life threatening symptoms.
- Prio 2: Urgent, not life threatening symptoms.
- Prio 3: Non-urgent calls.

Some counties also have a fourth degree for transportations where medical attention is not needed.

From the moment it has been determined that a person is in need of an ambulance, until a suitable resource has reached the patient, he or she is in the *SOS Waiting room.* This is not a physical place; the patients are still located at the call site. The waiting room is a way of keeping track of and managing patients that have not yet been reached by any ambulance. Tasks that are performed in the waiting room function may include calling back the patient at regular intervals or giving medical advice. While the patient is in the waiting room, the priority of the call may change; the symptoms may worsen, or the patient might start to feel better. The arrows to and from the ambulance planning and ambulance control function in Figure 1 indicate that what happens in the waiting room affects these functions.

Prio 1 calls are the most urgent, and the closest available ambulance is always sent to the call site, which is where the patient is located. The ambulance travels using siren and lights to get a clear path. When the ambulance personnel reach the patient, the call is *served*, or *executed.* To second priority calls, it is also stated that the closest available ambulance should be used. This is however interpreted differently in different counties; in some counties an ambulance can be regarded as not available if it is needed to cover an area for possible incoming calls. The ambulance is then maintaining the *preparedness* in the area.

In this case, an ambulance dispatcher may choose to assign an ambulance with a longer travel time to the Prio 2 call.

Prio 3 calls are often transportation of patients from their home to the hospital or from the hospital back home again, or between hospitals. They are non-urgent calls, and may have been received days before they are to be executed. If this is the case, the call is sent to the *Ambulance planning function*. Today, the planning performed in this function mainly consists of checking that all calls received in advance, which are referred to as *scheduled calls*, can be executed at the requested time. If this is not possible, a new time has to be suggested. It would be possible to perform more advanced transportation planning in the planning function, but this requires that a larger amount of non-urgent calls are received earlier than currently. Another reason for the sparse planning today is that a fixed transportation schedule would be disrupted by incoming urgent calls. One way of dealing with these disruptions is to dedicate a number of ambulances to scheduled transports, and thus not use them for urgent calls. Some counties in Sweden use this strategy. Another way is to ensure that the transportation schedule is robust and to use a re-planner to fix the schedule when it is disrupted, that is, execute some sort of disruption management (Clausen *et al* 2001).

Unscheduled calls consist of urgent calls where there is no time to make any transportation planning, and non-urgent calls where the patient still desires an ambulance as quickly as possible. The unscheduled calls are sent from the Order handling function to the *Ambulance control function*. It is interesting to note the difference in objectives between a scheduled call and an unscheduled call (see Figure 2).

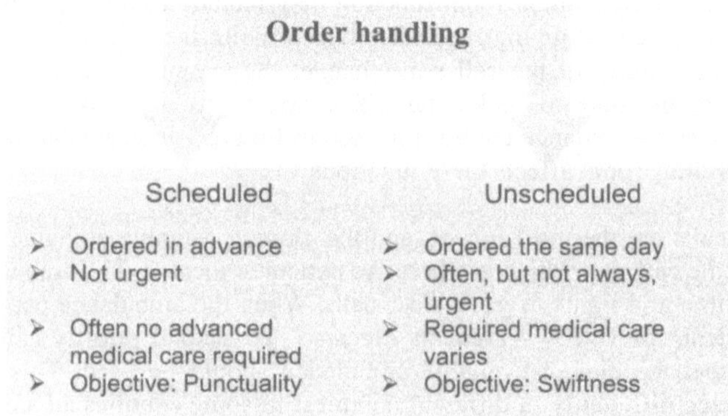

Figure 2: Characteristics of scheduled and unscheduled calls.

A scheduled call has an expected patient pick-up time, which is the time when the ambulance is supposed to be able to serve the patient. The objective for SOS Alarm in this case is clearly to make sure that the ambulance is not late, or even too early, since the resource cannot do anything useful while waiting for the patient. The objective for unscheduled calls is to get an ambulance to the patient as quickly as possible. Many counties in Sweden have patient waiting time targets, where the patient waiting time is the elapsed time from when the call was received by an SOS operator until the ambulance personnel have reached the patient. The county of Stockholm, which is the largest county in Sweden in terms of number of calls and ambulances, has different patient waiting time targets depending on the priority of the call. For Prio 1 calls, the target is that 75% of all patients should be served within 10 minutes. This means that the patient waiting time should be less than 10 minutes for 75% of the patients. Furthermore, 95% should be served within 15 minutes and 99% within 20 minutes. For Prio 2 and 3 patients, the service times are allowed to be longer. As far as the authors are aware, no county use any kind of quantifiable targets for scheduled calls.

All unscheduled calls are sent to the Ambulance control function, where an ambulance dispatcher (sometimes also referred to as an ambulance controller) tries to find a suitable resource to serve the call. When making this decision, the dispatcher has to regard the equipment of the available resources, the qualifications of the people manning the resources and the location of the call and of the available resources. As mentioned above, in the case of a Prio 1 call, the main objective is to get a resource to the patient as quickly as possible. This means sending the closest ambulance, but may also mean sending a non-qualified ambulance resource, or even a fire service resource, as a first measure if this is closer. The dispatcher may also re-assign an ambulance already on its way to a less urgent call to take the new one instead.

The fleet of vehicles varies from county to county. SOS Alarm does not own any ambulance resources. In some counties, the county council owns the ambulance resources and in other counties, the ambulances are owned and operated by independent suppliers. Mixes of these two alternatives occur as well. Whichever the case, it is the county councils who are primarily responsible for the health care in a county and thus also for the ambulance services. The types of ambulances can differ between counties. The regular resource, here simply called the ambulance, travels by road and can handle any Prio 1 – 3 call but only transport one patient. There also exist aircraft for longer transports, and helicopters for areas that are difficult to access by car (e.g. islands) or when the distance is too long for a car. Other resources include light-ambulances that can be used for non-urgent calls, where only simpler medical treatment is necessary, and urgency-cars that cannot transport any patient but have personnel with an advanced medical training. The county councils decide what kind of resources and how many should be available, as well as their initial locations.

The main computer system in an SOS centre is called CoordCom, originally developed by the Swedish company Ericsson in 1987. CoordCom supports communication by radio, telephone and data, and is used to book, edit and store information about ambulances and calls. Information about a new call might, for example, be called in through the public switched telephone network, in which case CoordCom automatically checks, and enters the address from where the call originated. The SOS operator prioritises the call and enters additional information into CoordCom, and CoordCom indicates for the ambulance dispatchers that there is a new call to be served. The ambulance dispatcher assigns an ambulance to the call, and the system sends a data message, using the Mobitex network, to the ambulance personnel. When the ambulance is on its way to the call site, the ambulance personnel acknowledges this by sending a data message back to the centre. The Mobitex network is also used to transmit the position of the ambulance, which is received via the global positioning system (GPS), at regular intervals. A new system, Zenit, that will replace CoordCom, is being developed by Ericsson, and has already been taken into service in some centres.

The ambulance dispatchers also have a geographical information system (GIS) to support them in their work. The old GIS was recently replaced by the ResQMap system, developed by the Swedish company, Carmenta. ResQMap is connected to CoordCom and can graphically display the information that is stored in CoordCom, including the location of calls and ambulances.

Areas of Improvement

Taking the functions (see Figure 1) in SOS Alarm's ambulance logistics service as a starting point, it is possible to identify a number of areas of improvement. Some of these areas were already identified within SOS Alarm when this project started, while others came up as a result of the project work. In Figure 3, the areas have been placed directly below the function or functions that will be affected by work done in the area. Successful work in any area will be likely to improve the ambulance logistics service.

The areas are not as disparate as Figure 3 indicates though; work done in one area will have an impact on ongoing work in other areas of improvement, and in some cases it may be necessary to complete work in one area to be able to lucratively work in another. For example, in order to develop and implement efficient decision support tools, which is one objective in area 7, it is essential that there is relevant statistical information, i.e. work is carried out in area 2.

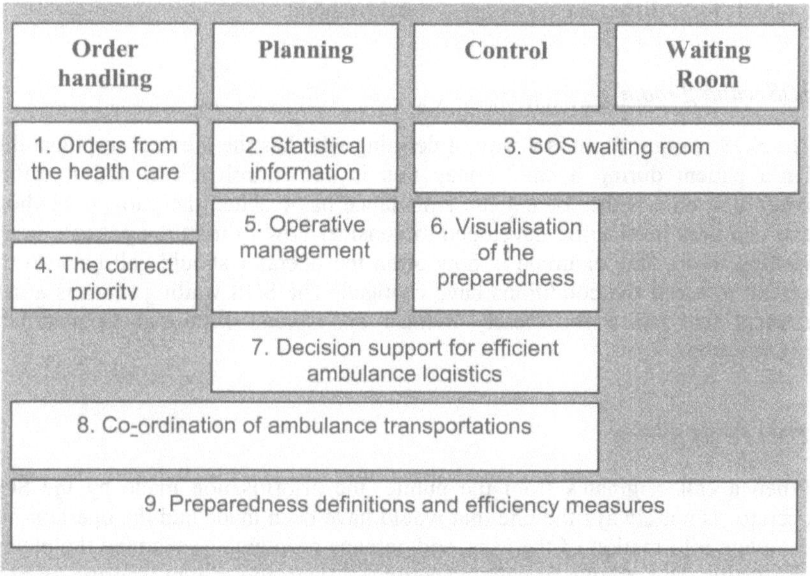

Figure 3: Areas where the ambulance logistics service can be improved

Orders from the health care

When a call is received from a hospital or another health care unit, the person who is making the request often states how urgent it is. This will then be a base for the prioritisation, which is done by an SOS operator. Currently, calls are sometimes made out to be more urgent than they really are. This may lead to calls being too highly prioritised, which means that the ambulance resources are not used in the most effective manner. The objective in this area is to find ways to ensure that each patient gets the proper type of transportation depending on his or her specific needs. The main work in this area involves obtaining better orders from health care personnel.

Statistical information

Today, statistical information, for example regarding when and where calls are likely to appear, is rarely used to support decisions within SOS Alarm. This is mainly due to the inability of the technical systems used in the SOS centres to supply relevant statistical data. As new technical systems are currently being developed and installed, this should not be a problem in the future. Still, methods for utilising the data have to be developed.

SOS waiting room

The SOS waiting room is a way of defining who has the medical responsibility for a patient during a call. Today this is not entirely clear, and it differs depending on whether or not the ambulance has reached the patient. Methods and practices need to be developed to consider how to treat the patients in the waiting room. One example is how often the operator should call back to the patient to see if the conditions have changed. The SOS waiting room is a new concept that has to be clearly defined and spread within and outside the organisation.

The correct priority

When a call originates from the public, the prioritisation made by the SOS operator is not always the one that would have been made had the operator had complete information of the case, and perhaps been able to examine the patient physically. This results in calls sometimes having too a high priority, as SOS Alarm does not risk assigning too low a priority to a call to ensure the safety of the patient. As often as possible, the priority of the call should be exactly as required from the state of the patient. The main difference between this area and area 1, is that here the prioritisation usually has to be made based on information from people who are not medically trained. Therefore, the work in the area might include the development of more effective interviewing techniques.

Operative management

Any disaster will require massive ambulance resources, and often fire service and police resources are utilised as well. In order to manage and control these, a management function is set up at the disaster area. The efficiency of this management function is directly dependent on the communication and co-operation with other management functions, including the SOS centres. It is therefore vital that this collaboration is functioning in a satisfactory manner.

Visualisation of the preparedness

The GIS (geographical information system) used in the SOS centres is capable of showing the positions of calls and ambulances. This is of great help to the ambulance dispatchers. The GIS could be of even more assistance if it was possible to visualise the preparedness directly in the GIS, for example by colour-coding areas with low levels of preparedness.

Decision support for efficient ambulance logistics

Even if the ambulance dispatchers are skilled and experienced, appropriately designed decision support tools could help them find better solutions or assist them in routine matters. Automatic decision support can be of use in all parts of the planning process, that is, in strategic and tactical planning as well as in operational control. In the strategic planning, a decision support tool may help evaluate strategic decisions, such as the sizing of the fleet or the location of ambulance stations. Tactical support may include routing of ambulances for non-urgent calls, and in the operational control, a decision support tool can suggest which ambulance to assign to an incoming call. A human ambulance dispatcher is still needed to evaluate the suggestions and make the final decisions.

Co-ordination of ambulance transportations

The planning at a hospital, for example in which order patients are to be treated, is commonly made without regarding the transportations to and from the hospital. This means that sometimes a patient living far away from the hospital will be scheduled for treatment early in the morning, and thus has to be transported to the hospital the evening before, and stay the night. If ambulances are needed to perform these transportations, the planning made at the hospital also might make it impossible to construct an effective transportation schedule. A more resource-effective way of transporting the patients could be achieved if the planning at the hospital was co-ordinated with the planning of the ambulance resources.

A related issue is the planning of ambulances that operate outside their own county. From a system wide perspective, it would be advantageous if ambulances could service patients without regard to which county they belong to. One obstacle for this kind of co-ordination is the complicated rules that the county councils sometimes set for using each other's ambulances. Another complication comes from the fact that different organisations are responsible for different types of patient transportations, depending on which kind of resource is needed, e.g. ambulance, taxi or aircraft.

Preparedness definitions and efficiency measures

Today, a unitary set of definitions and measures for describing and evaluating the quality of SOS Alarm's ambulance logistics service is lacking. To be able to review the effects of different actions and projects, such a set is necessary. It would also make it easier for SOS Alarm to describe their business, and in a compact way supply their customers with quality reports and decision support.

Decision Support Tools

In this section, work within the areas of improvement 6 and 7 (see Figure 3) is described. The work concerns the development of a number of decision support tools, primarily for the ambulance dispatchers. The first tool, called the preparedness calculator, is directly related to area 6. However, the tool can also be considered a result from work in area 7, where the development of decision support for ambulance logistics is the main objective. Three other tools are presented: a dispatch tool, a relocation tool and a simulation tool.

A preparedness calculator

To state the preparedness in ambulance logistics is a way of describing the current operational situation. If the preparedness is "good", the ambulance dispatcher feels confident that it is possible to serve effectively a normal amount of calls in the near future. This situation might however change rapidly; suppose that there is a serious traffic accident, forcing the dispatcher to send a large number of ambulances to the call site. These ambulances will probably be the ones that are closest to the call site, making the surrounding area devoid of available ambulances. Suddenly the preparedness is "bad", at least around the call site. Naturally, "good" and "bad" are quite vague expressions, and do not give a clear picture of the operational situation.

In order to find a quantifiable measure for preparedness, the area of responsibility, e.g. a county, is first divided into a number of different zones. This makes it possible to calculate the preparedness for different parts of the county, that is, the level of preparedness can be adequate in one zone, while it is low in another. The preparedness depends on how many ambulances that can reach a zone, and how far away from the zone they are. It is also affected by the expected need for ambulances in the zone, which depends on the size and age of the population, the criminality rate, the traffic intensity and many other factors.

Given a certain fleet of ambulances with a known status and location, the preparedness in each of the zones can be calculated. The most obvious area of use for this is in the operational control, where the preparedness can be updated dynamically. The ambulance dispatcher can check when the preparedness is low in a zone and do something to correct it, for example, relocate an ambulance to cover the affected zone.

To benefit fully from the preparedness measure, it should be presented in a way that makes it easy for the ambulance dispatcher to evaluate the situation. This can be implemented by using different colours to represent different levels of preparedness in the GIS.

Before the preparedness measure can be used, it has to be calibrated, verified and evaluated. The calibration phase includes collecting necessary data and finding parameter values. For the calculation of the preparedness, this means constructing the zones, finding expected demands for ambulances and obtaining expected travel times between the zones. Each one of these tasks is a research challenge in itself, but fortunately this also means that there often has been some prior work in the area (see e.g. Goldberg, 2004).

To calibrate the preparedness measure for the county of Stockholm, travel time data has been acquired for 1240 zones. The data consists of expected travel times between each pair of zones in the county, and was provided by Inregia AB. This data is static; for instance, the travel time between two zones is the same in the morning as in the evening. For the measure to be truly useful in an operational setting, more than one set of travel times has to be obtained, preferably one set for each distinctive traffic situation. The ultimate goal is to be able to incorporate real time traffic information about expected travel times, as well as traffic jams and accidents, into the systems in the SOS centre.

To calculate the expected demand for ambulances in the zones, historic call data is used as a base. Data for one year is collected for the different ambulance districts in the county of Stockholm. Using this data, it is possible to construct a forecast for each ambulance district, for each hour of the week. This means that a certain district will have a specific number of expected calls depending on the day and the hour. Since the ambulance districts are much larger than the zones mentioned above, population data for the zones is used to distribute the demand for one district over the zones included in the district. The available population data consists of the number of people living in the zone, how many of these have employment, and the number of people that have their place of work in the zone. This makes it possible to calculate different populations in a zone for daytime and nightime, which further refines the forecasts.

An ambulance dispatch suggestion tool

The main decision in ambulance dispatch is to choose which ambulance to assign to each call. Sometimes this is easy; for a Prio 1 call that requires only one ambulance, the ambulance with the shortest expected travel time to the call site is always dispatched. However, the dispatcher must still ensure that the ambulance carries the necessary equipment and that the ambulance personnel are qualified to handle the call. When the call is not as urgent, an ambulance dispatcher may choose to dispatch an ambulance with a longer travel time, if this assignment means that the drop in preparedness will be less significant. The dispatcher may also re-assign an ambulance already on its way to a call site, if the new call is more urgent.

The most common and natural dispatch rule is to send the closest unit, since the general objective is to minimise the response times. However, Carter *et al* (1972) show that this rule is not always optimal. They consider a case where two units, A and B, have equally large areas of responsibility, but A's area has a significantly higher call frequency. In this case, the mean response time will decrease if B is allowed to respond to some of the calls for which A is the closest unit. This result was generalised in Cunningham-Green and Harris (1988) for cases involving more than two units. Repede and Bernardo (1994) consider that it may be better to send unit C to take A's call when A is busy, than to send the closer unit B. This is done if the call frequency in B's primary district is higher than in C's. Weintraub *et al* (1999) suggest a dispatch support system for vehicles servicing the electrical system in Santiago de Chile. Vehicles travel from call site to call site, and the dispatcher tries to maintain an adequate preparedness for quickly servicing high priority calls when deciding which unit should be assigned to each call.

The dispatch tool developed for SOS Alarm automatically allocates an ambulance to a new incoming call. The preparedness calculator keeps track of which ambulances are close to a zone, which makes it easy to find the closest ambulance to a certain zone. For a Prio 1 call, the closest ambulance is always dispatched. To check which ambulance to dispatch to a Prio 2 or 3 call, the tool checks all available ambulances within a certain travel time from the zone, and picks the one whose unavailability causes the least drop in the preparedness. This means that the dispatch tool sometimes assigns an ambulance that is not the closest one to the patient, if the call is not Prio 1. However, this can only happen if there is more than one ambulance available within a reasonable time. This time can, for example, be set in accordance to the patient waiting time targets, which in the case of Stockholm means that if there is an ambulance within 30 minutes from a Prio 2 call site, the patient will get an ambulance within that time.

The tool also re-assigns ambulances already on their way to serve a call, to calls that are more urgent; for instance, an ambulance on its way to a Prio 3 call, can be assigned to a new Prio 2 or Prio 1 call.

An ambulance relocation tool

An ambulance dispatcher may relocate an ambulance if they believe that there is a location where the ambulance is more likely to be close to a new call. This relocation will then increase the preparedness.

A dynamic relocation algorithm for fire companies was developed by Kolesar and Walker (1974), and a call for relocation is triggered when some part of the city is not covered by any unit. Gendreau *et al* (2001) present a tabu search heuristic for the dynamic relocation of ambulances and a model for physician cars is presented in Gendreau *et al* (2005). Both models maximise the coverage of the area.

The relocation tool described in this paper simulates a relocation decision by assuming that relocating one or more ambulances is beneficial if the level of preparedness, as given by the preparedness calculator, drops below a certain threshold value, P_{min}. This gives rise to the following optimisation problem:

Minimise

{the maximum relocation time among the relocated ambulances}

subject to

{all zones must have a preparedness level of at least P_{min} after the relocation is complete}

{at most R_{max} ambulances may be relocated}

Today, the ambulance dispatchers solve the problem above manually, although the objective and the constraints are not quantified. In order to simulate the relocation decisions, quantifications are necessary. The decision to solve the problem is triggered by any zone having a level of preparedness below P_{min}, and the objective is to correct this situation as quickly as possible. The constraint that not more than R_{max} ambulances may be relocated is necessary to ensure that it will be possible to use the solutions to the problem in practice; no ambulance dispatcher would approve of a solution that involved too many relocations. If more than two or three ambulances are relocated, it is hard for the dispatcher to evaluate the solution, and might feel that it is not worth the trouble to execute it. A greedy tree search heuristic is used to find solutions to the relocation model. Today, the model may relocate an ambulance to practically any part of the area of responsibility. It is however possible to limit the feasible relocation sites, which will shorten the computation time for the heuristic.

An ambulance simulation tool

Simulation has been used extensively in ambulance logistics research, foremost as a way of evaluating different locations of ambulance stations. One early simulation model for evaluating possible improvements in ambulance service is described in Savas (1969), and a more recent simulation study is described in Henderson and Mason (2004). Furthermore, the hypercube model (Larson, 1974), and later extensions of this work, can be used to evaluate a solution from a location model.

A simulation of an ambulance services system includes patients, ambulances and ambulance dispatchers. The patients are foremost characterised by their location, their possible destination and the urgency of their situation. Depending on the desired level of detail or direction of the study, it is possible to add more patient properties. Ambulances are used for medical treatment and the transportation of the patients. They are characterised by their position and their status, which for example can be *available*, *en route to Prio 1 call*, or on a *lunch break*. The simulator has to find routes for the ambulances when needed, for example when they are dispatched to serve a call.

A model capable of simulating patients (calls) and ambulances can be used as an educational tool. An ambulance dispatcher will then make all the decisions, such as the assignment of calls to ambulances and the dynamic relocation of ambulances. When using simulation to evaluate more strategic questions, for example about fleet sizing and ambulance station location, it is necessary to simulate the ambulance dispatcher as well. To get significant results that can answer the strategic questions, many hours of operations have to be simulated, and using a human ambulance dispatcher for this is too time consuming. Instead, by using the dispatch and the relocation tools, the decisions made by an ambulance dispatcher can be simulated, and the simulations can be run faster than real time.

Today, the GIS in an SOS centre is able to, among other things, visualise and manage information about calls and ambulances. This makes it an ideal platform for an ambulance control simulator, especially since the ambulance dispatchers are already used to the system interface.

In order to simulate calls, a call generator emulating the information that is usually received from CoordCom is needed. Calls can be generated stochastically according to historically verified distributions, which is proper if the simulator is to be used for evaluating strategic decisions. A constructed sequence of calls may be a preferred input if the simulator is used as an educational tool, since this makes it possible to build scenarios to educate and test the skills of the ambulance dispatchers. The constructed sequence may also be a set of historical calls.

The simulation tool that has been developed so far is much simpler than what could be achieved by an implementation in the GIS. It is capable of generating random calls, where a high population in a zone gives rise to a higher call frequency. The dispatcher tool can assign ambulances to the calls, and the relocation tool finds relocations when the level of preparedness in a zone drops below the threshold value. When assigned to a call, the ambulances travel from their stations to the call site, pick up the patient and travel to the closest hospital. When an ambulance is finished with the patient, it returns to the ambulance station, unless it has already been assigned a new call or been given a relocation order.

Simulations have been run to evaluate the other decision support tools (Andersson, 2005). The results show that the relocation tool has a positive effect on the patient waiting times. This indicates that trying to maintain a high preparedness as given by the preparedness calculator, also reduces the patient waiting times, which verifies that the preparedness calculator gives an accurate picture of what the ambulance dispatchers refer to as preparedness.

Final Remarks

Currently SOS Alarm collaborates with Carmenta and Linköping University in a project to install the decision support tools as modules to the GIS in the SOS centres in Stockholm and Falun. Computational testing and theoretical reasoning has shown us the benefits of the new applications, but it is not until the intended users have evaluated them, that their true worth will be clear.

References

Andersson T (2005). *Decision support for dynamic fleet management.* PhD thesis, Department of Science and Technology, Linköping University, Sweden.

Carter G, Chaiken J and Ignall E (1972). Response areas for two emergency units. *Operations Research.* 20(3): 571–594.

Clausen J, Hansen J, Larsen J and Larsen A (2001). Disruption management, *OR/MS Today*, 28: 40–43.

Cunningham-Green R and Harries G (1988). Nearest-neighbour rules for emergency services. *ZOR – Zeitschrift for Operations Research.* 32: 299–306.

Gendreau M, Laporte G and Semet S (2001). A dynamic model and parallel tabu search heuristic for real-time ambulance relocation. *Parallel Computing.* 27: 1641–1653.

Gendreau M, Laporte G and Semet S (2005). The maximal expected coverage relocation problem for emergency vehicles. *Journal of the Operational Research Society.* 57(1): 22–28.

Goldberg J (2004). Operations research models for the deployment of emergency service vehicles. *EMS Management Journal.* 1: 20–39.

Henderson S and Mason A (2004). Ambulance service planning: Simulation and data visualisation. In Brandeau M, Sainfort F and Pierskalla W (eds). *Operations Research and Health Care.* Kluwer Academic Publishers, Boston. pp 77–102.

Kolesar P and Walker W (1974). An algorithm for the dynamic relocation of fire companies. *Operations Research.* 22(2): 249–274.

Larson R (1974). A hypercube queuing model for facility location and redistricting in urban emergency services. *Computers and Operations Research.* 1(1): 67–95.

Repede J and Bernardo J (1994). Developing and validating a decision support system for locating emergency medical vehicles in Louisville, Kentucky. *European Journal of Operational Research.* 75: 567–581.

Savas E (1969). Simulation and cost-effectiveness analysis of New York's emergency ambulance service. *Management Science.* 15(12): B608–B627.

Socialstyrelsen (2001). Ambulanssjukvårdens termer och begreppsdefinitioner, Riktlinjer.

Svenskt index för akutmedicinsk larmmottagning (2001), Svenska läkaresällskapet.

Weintraub A, Aboud J, Fernandez C, Laporte G and Ramirez E (1999). An emergency vehicle dispatching system for an electric utility in Chile. *Journal of the Operational Research Society.* 50: 690–696.

MODELLING THE ASSIGNMENT OF OUTPATIENT EXAMINATION ROOMS

STEVE GALLIVAN
Clinical Operational Research Unit, Department of Mathematics, University College London, UK

Abstract

A large London teaching hospital has recently rebuilt the main hospital building and is reorganising its outpatient services, reducing the number of outpatient examination rooms from 90 to 55. They faced the problem of assigning rooms to clinical specialities, given that the number of clinics for each speciality varies from session to session of the week. Also, some examination rooms are more suited to some specialities than others. It was found that this problem could be expressed in relatively simple terms in integer programming terms. Further, the particular form of the problem meant that integer solutions occurred automatically if one merely solved the associated linear programming problem. In view of this, a very simple prototype planning tool could be constructed using the facilities of Excel, avoiding the need for specialist mathematical programming packages.

Keywords: scheduling; optimization; planning.

Introduction

This project arose as a result of a request from a large teaching hospital concerning the planning of outpatient clinic sessions in a new facility that was being built for outpatients. There were some 15 broad clinical specialities and the new facility had 55 examination rooms whereas, prior to the new building, some 90 examination rooms had been used, which was thought to represent considerable over-provision.

This is a complex planning problem since there are a huge number of potential combinations of room assignments and potential schedules. On the face of it, there are several operational problems notably:

- varying patterns of availability of clinicians during the week, due to other commitments;

- inappropriateness of some rooms for some clinical specialities.

Both of these problems had some elements that imposed constraints that were very difficult to avoid, (e.g. some clinic types require a room equipped with specialist equipment), while others represented somewhat softer constraints, (e.g. Sir Lancelot Spratt prefers to operate elsewhere on a Friday afternoon).

Estimates of problem complexity are alarming. Assuming a planning cycle of one week, each with 10 sessions, there would be 2^{285} different potential time-tables assigning clinics to specialities and rooms, a number bigger than the number of atoms in the universe. Even if the clinical specialities requiring rooms for a given day were known, there are still 551 different potential room assignments. Even with modern day computers, evaluation of each possibility separately is not a feasible option. It was the latter problem that the hospital management was most concerned about.

This paper stems from a feasibility study carried out to examine the scope for applying mathematical programming techniques to this problem. Initial investigation soon identified a way in which an integer programming formulation could be structured (Gallivan and Utley, 2004). Applying mathematical programming in the context of hospital planning and operation is certainly not original (Blake and Donald, 2002; Blake and Carter 2003; Adan and Vissers, 2002). However here, the form of the integer programming formulation that was devised involved many decision variables and would thus require large scale computing requirements and specialist software which the hospital had neither the expertise to use nor the wish to buy without evidence. Further analysis produced a modified version of the model that greatly reduced the number of variables required. Further, the modified model expressed the problem in a form that has a special form of mathematical symmetry. This enabled pure mathematical techniques to be applied, establishing that simple linear programming methods would automatically generate integer solutions for such a problem.

As a result, a prototype software tool for assisting planning was developed that makes use of the solver facility of Excel, which is well known by the hospital planners.

Assigning Clinic Rooms

It may be that some clinic-room matches are either inconvenient or judged to be virtually impossible. For example, clinics attended by patients with mobility problems should ideally be matched to rooms near the entrance, while some clinics may require special equipment and only be matched to a room that is suitably equipped.

Mathematically, this is treated in two ways. First of all, there are hard constraints that must be observed, which either force particular matches to be made or avoid others being considered.

On the other hand, preferences (rather than musts) are treated by pre-specifying a notional cost or tariff associated with each potential matching option – high costs corresponding to options that should be avoided, if possible.

Specifying resources and planning needs in terms of mathematical notation

In this section we define mathematical notation to represent planning needs in quantitative terms. In computational terms, these quantities are regarded as constant input data describing the system.

Assume that the available rooms have been grouped into S different room groups each comprising rooms with broadly similar characteristics in terms of location and specialised equipment.

For $1 \leq s \leq S$, assume that at most R_s examination rooms are available for the s-th room group.

Assume that these rooms must be matched to clinics for M clinical specialities. For $1 \leq m \leq M$, let the integer T_m denote the minimum number of clinics that the m-th clinical team must have during the planning cycle.

For $1 \leq s \leq S$, $1 \leq m \leq M$, let $C_{s,m}$ denote the notional cost associated with matching a clinic for the m-th clinical team to the s-th room-group. It is assumed that these costs are all non-negative, large notional costs being associated with inconvenient assignments.

Finally, to reflect the fact that some assignments may be regarded as absolutely necessary and some should be avoided at all costs, we define upper and lower bounds for the assignments that can be made.

For $1 \leq s \leq S$, $1 \leq m \leq M$, let $d_{s,m}$ denote the minimum number of clinics from room-group s that should be assigned to clinics from speciality m, and let $D_{s,m}$ denote the maximum number of clinics from room-group s that should be assigned to clinics from speciality m.

Decision variables

The quantities described in the previous section are regarded as fixed in respect of the planning process. Here we describe the variables whose values need to be determined.

For $1 \leq s \leq S$ and $1 \leq m \leq M$, we define integer variables $\{X_{s,m}\}$ such that $X_{s,m}$ is the number of rooms assigned to the m-th clinical team from the s-th room-group.

Framing the optimisation problem

The planning process effectively involved choosing the decision variables $\{X_{s,m}\}$ in such a way as to best meet a goal. However in doing this there are some constraints that govern the choices that can be made. The particular way that variables have been defined has been chosen to make it straightforward to specify this in simple mathematical terms.

The first constraint is that one cannot assign more rooms than are available. This can be expressed in the following terms:

$$\sum_{m=1}^{M} X_{s,m} \leq R_s, \qquad 1 \leq s \leq S. \tag{1}$$

The next constraint ensures that the number of clinics assigned to each clinical team is sufficient to cater for their needs.

$$\sum_{s=1}^{S} X_{s,m} \geq T_m, \qquad 1 \leq m \leq M. \tag{2}$$

Optimal matching of rooms to clinics that provides the minimal notional cost while observing all the hard constraints can be obtained by solving the following integer programming problem.

Minimise $\sum_{s=1}^{S}\sum_{m=1}^{M} C_{s,\ m}X_{s,\ m}$

Subject to:

$\sum_{m=1}^{M} X_{s,\ m} \leq R_s,$	$1 \leq s \leq S.$	(3)
$\sum_{s=1}^{S} X_{s,m} \geq T_m,$	$1 \leq m \leq M;$	(4)
$d_{s,m} \leq X_{s,\ m} \leq D_{s,m}$	$1 \leq s \leq S,\ 1 \leq m \leq M;$	(5)
$X_{s,\ m}$ an integer	$1 \leq s \leq S,\ 1 \leq m \leq M.$	(6)

It should be noted that for all its apparent complexity, there is a pleasing symmetry to this, at least to the eyes of mathematicians. It expresses the problem with variables that correspond to elements of a matrix and has similarities to a standard transportation problem, but with range constraints on the matrix elements. Further, the only other constraints involve row and column sums.

For the particular teaching hospital being considered, the number of decision variables is of the order of 200, so this is a computationally quite a complex problem. However, with great good fortune, constraint (6) is not necessary. The symmetry of constraints (3), (4) and (5) ensures that so long as their bounding constants are integers (which are the only sensible values) then treating the problem as a linear programming problem automatically generates integer solutions. This enables the problem to be solved in a matter of seconds.

No claim is made that this integrality result is original, it would be curious if it were. but the author was certainly unaware of it.. Providing proof for this curious property required the author to apply very rusty pure mathematical skills rarely used since his PhD in 1975 in the field of convexity. Details of the proof are reported elsewhere.

Practical Implementation of the Method

A prototype spreadsheet planning tool was developed, based on the modelling described above. This has the acronym POPS (**P**rototype **O**utpatient **S**cheduler) and makes use of the "Solver" facility of Excel which allows linear programming to be performed, although it is limited to 200 decision variables. By coincidence, 200 is exactly the number of decision variables used by POPS. POPS operates with a number of linked worksheets.

The user has access to several worksheets and fictitious data is used to illustrate the functioning of the system. Typically only one of these contains data that might change from session to session , i.e. specifying which doctors have booked clinics. Other worksheets, once set, would rarely be changed. These specify things such as the number of room groups, the number of rooms in each and the notional tariff associated with matching a particular speciality with a particular room group. Data in these worksheets is used to construct arrays on a 'calculation' worksheet that specify constraint sets (3) and (5) and the coefficients of the cost function in the linear programming formulation described in the previous section.

A worksheet that might change from session to session is where the user inputs the basic data about the clinics that need to take place during the planned session. An extract from data that might typically be held in this worksheet is shown in Table 1. The data corresponding to columns 3 and 4 of Table 1 is used to create an array on a separate worksheet that specifies the constraint set (4).

When the POPS system is run one of two things will happen. The system may report that no feasible solution could be found. This means that, given the information about hard constraints (room availability, clinic numbers and clinician availability), there is no schedule that satisfies all these constraints.

Alternatively, within a second or two, the system reports that it can find a solution and the schedule that provides the minimum total inconvenience tariff is displayed. Typical output is to an Excel worksheet as summarised in Table 2.

Consultant	Clinic name	Speciality/ Clinic type	Clinic booked	Patients booked
Mr Hookes	Prosthetics	General surgery A	1	30
Mr Key	Gall bladder	General surgery A	1	29
Mr Green	ENT	General surgery B	0	0
Mr Hammond	Organ transplant	General surgery B	1	20
Mr Wessels	Vascular surgery	General surgery B	0	0
Mr Duff	Obs & Gynae	General surgery B	1	22
Mr Wombwell	Obs & Gynae	General surgery B	0	0
Mr Sidebottom	Stoma clinic	General surgery C	0	0
Mr I.Ball	Cataract extraction	Ophthalmology	1	30
Dr Knott	Re-operations	General medicine A	1	30
Dr Blowers	Respiratory medicine	General medicine A	0	0
Dr Meandad	Genetic diseases	General medicine A	1	20
Dr Hick	Alcohol abuse	General medicine A	0	0
Dr Sobers	Alcohol abuse	General medicine A	0	0
Dr Lush	Alcohol abuse	General medicine A	1	19
Dr Snow	Drug abuse	General medicine A	1	18
Dr Ashe	Smoking cessation	General medicine A	0	0
Dr Fagg	Smoking cessation	General medicine A	1	21
Dr Old	Geriatrics	General medicine A	1	20
Dr Young	Paediatrics	General medicine A	0	0
Dr Child	Paediatrics	General medicine A	0	0
Prof Willy	Urology	Urology	0	0
Prof Peters	Urology	Urology	1	15
Mr Cox	Urology	Urology	1	15

Table 1: Typical data (fictitious) specifying clinicians who have clinics booked for a particular session, their speciality and the number of patient examinations involved.

Specialty	Recommended number of clinic sessions to assign to each speciality for each time slot						Total number of clinics recom-mended	Num-ber of clinics needed
	Room group 1	Room group 2	Room group 3	Room group 4	Room group 5	Room group 6		
GENERAL SURG. A	0.0	0.0	2.0	0.0	0.0	0.0	**2.0**	2.00
GENERAL SURG. B	0.0	0.0	0.0	1.0	0.0	1.0	**2.0**	2.00
GENERAL SURG. C	0.0	0.0	0.0	1.0	0.0	0.0	**1.0**	1.00
OPHTHALMOL. A	0.0	0.0	0.0	0.0	1.0	1.0	**2.0**	2.00
OPHTHALMOL. B	0.0	0.0	0.0	0.0	0.0	0.0	**0.0**	0.00
GENERAL MED. A	1.0	0.0	0.0	0.0	4.0	1.0	**6.0**	6.00
GENERAL MED. B	0.0	1.0	1.0	2.0	1.0	0.0	**5.0**	5.00
GENERAL MED. C	3.0	0.0	2.0	0.0	0.0	0.0	**5.0**	5.00
UROLOGY	0.0	4.0	2.0	0.0	0.0	0.0	**6.0**	6.00
TRAUMA	0.0	0.0	2.0	0.0	0.0	0.0	**2.0**	2.00
DIABETES	0.0	0.0	0.0	0.0	0.0	0.0	**0.0**	0.00
PLASTIC SURGERY	0.0	0.0	0.0	1.0	0.0	2.0	**3.0**	3.00
BREAST	0.0	0.0	0.0	0.0	0.0	1.0	**1.0**	1.00
PRE-ASSESSMENT	0.0	0.0	0.0	2.0	0.0	0.0	**2.0**	2.00
GUT	0.0	1.0	0.0	0.0	0.0	0.0	**1.0**	1.00
GYNAECOLOGY	0.0	0.0	1.0	0.0	0.0	0.0	**1.0**	1.00
PHARMACOLOGY	1.0	1.0	0.0	0.0	0.0	0.0	**2.0**	2.00
CARDIOLOGY	0.0	2.0	0.0	0.0	0.0	0.0	**2.0**	2.00
RENAL	0.0	0.0	0.0	0.0	0.0	1.0	**1.0**	1.00
DERMATOLOGY	0.0	0.0	1.0	0.0	0.0	0.0	**1.0**	1.00
RHEUMATOLOGY	0.0	0.0	0.0	0.0	1.0	0.0	**1.0**	1.00
INFECTIONS	0.0	0.0	0.0	1.0	0.0	0.0	**1.0**	1.00
ENDOCRINIOLOGY	0.0	1.0	0.0	0.0	0.0	0.0	**1.0**	1.00
THORACIC	3.0	0.0	0.0	0.0	0.0	0.0	**3.0**	3.00
CLINICAL PHARM.	0.0	0.0	0.0	0.0	1.0	0.0	**1.0**	1.00
DIET A	1.0	0.0	0.0	1.0	0.0	0.0	**2.0**	2.00
DIET B	0.0	0.0	0.0	0.0	0.0	1.0	**1.0**	1.00
Total recommended	**9.0**	**10.0**	**11.0**	**9.0**	**8.0**	**8.0**		
Rooms available	9.0	10.0	11.0	9.0	8.0	8.0		

Table 2: An example of optimised room assignment minimising total inconvenience tariff calculated by POPS system (fictitious data)

POPS recommendations are by no means perfect and should not be regarded as sacrosanct. They are a first step in the process, which first establishes whether the system is overloaded then, if not, provides some initial recommendations, which need to be scrutinised to see if they are sensible. It is up to the planner to decide issues such as whether it is sensible to have a room group with very few clinics scheduled. Given how fast the planner works it can be used as a tool to do 'what if' analysis, using the various constraints as a means of forcing particular assignments. For example, a room-group might be judged to be under-used while another is too busy. Data in the constraint worksheets can easily be adjusted by hand to force more clinics to take place in underused room-group.

Discussion

This study established the feasibility of using an integer and linear programming approach to the problem of determining assignments of examination rooms to different clinics. A previous paper (Gallivan, 2005) also showed that there was scope for applying such methods to longer term scheduling issues to achieve a basic weekly clinic schedule, which takes account of clinicians' availability during the week.

References

Gallivan S, Utley M (2004) 'Modelling the assignment of outpatient examination rooms – what is the capacity of an outpatient department?', Clinical Operational Research Unit Discussion paper 695.

Blake JT, Donald J (2002), 'Using Integer Programming to Allocate Operating Room Time at Mount Sinai Hospital' *Interfaces*, 32(2): 63–73.

Blake JT, Carter MW (2002), 'A Goal-Programming Approach to Resource Allocation in Acute-Care Hospitals' *European Journal of Operational Research,* 140(3): 541–561.

Adan, I.J.B.F., & Vissers, J.M.H. (2002) 'Patient mix optimisation in hospital admission planning: a case study', *International Journal of Operations and Production Management, 22(4)*, 445–461.

Gallivan S (2005), 'Modelling the assignment and scheduling of outpatient examination rooms', Clinical Operational Research Unit, Report 729, February 2005.

OPTIMIZATION MODELLING OF HOSPITAL OPERATING ROOM PLANNING: ANALYZING STRATEGIES AND PROBLEM SETTINGS

MARIE PERSSON and JAN A. PERSSON
School of Engineering, Blekinge Institute of Technology, Ronneby, Sweden

Abstract

There is a growing proportion of elderly which increases the demand for health care. As a consequence health care costs are rising and the need for hospital resource planning seems urgent. Different aspects (often conflicting) such as patient demand, clinical need and political ambitions must be considered. In this paper we propose a model for analyzing a hospital surgical suite with focus on operating room planning. An optimization model is developed for patient operation scheduling and for key resource allocation. Medical examinations and treatments of patients are performed using a number of resources, similar to products being refined in a number of processes in a logistics chain. Optimal resource allocation, given different objectives according to patient perspective, staff perspective, costs etc. under different system settings (e.g. principles for operating room allocation and amount of stand-by personnel), is studied. Preliminary results are presented based on case studies from two Swedish hospitals.

Keywords: hospital; health service; integer programming; logistics; optimization; scheduling.

Introduction

At a hospital surgical suite several different resources have to be synchronized in order to achieve efficiency. Through the last decades, many studies have been conducted in an attempt to optimise health care delivery, see comments by Lagergren (1998), Riley (1999) and De Angelis *et al* (2003). Scheduling staff, beds and operating room allocations are common optimisation problems that are studied in this field (Beaulieu, 2000, Kua *et al,* 2003, Ogulata and Erol, 2003). Various approaches, including optimisation techniques, are used depending on how the problem is formulated and possibly separated into different parts. One of the main interests of optimizing health care is how the allocation of the operating room is arrived at. Previous research focusing on optimizing the surgical suite, generally proposed strategies to minimize monetary costs and to achieve as high a patient through-put as possible (Lowery, 1992, Hollingsworth, 2003). In this paper, we view operating room planning mainly from the patient perspective, which is not the typical view of this planning task. We also consider personnel perspectives and financial aspects.

Operating room planning is a complex task which has to consider many aspects such as surgeon scheduling, operating team scheduling (included anaesthetics personnel), patient related information, (i.e. estimated operating time, priority and diagnosis), equipment and related activities such as the intensive care unit etc. We suggest optimization modelling to support the allocation of key resources for operating room planning. The purpose is to show that this method can serve as a tractable tool for tactical planning within this environment. The case study is based on actual data from Swedish hospitals.

In this paper, we first introduce the problem as experienced by two hospitals of different types. Then an optimization model is suggested and the generation of input-parameters is presented. Different scenarios are created and the results produced by the optimization model are presented. Finally, the conclusions and directions for future research are outlined.

Problem Description: Operating Room Planning

Due to the complexity of operating room planning and the way surgery activity affects and is affected by many processes within hospital, it is a challenge to find out and analyze the parts which primarily influence the surgical activity. In this case study we try to identify primary requirements that form the basis of operating room planning. In order to obtain an understanding of the surgical activity from a general point of view, we have studied two cases at two Swedish hospitals;

- Blekinge Hospital (Blekingesjukhuset)
  A medium sized hospital (ca. 420 beds).
  Department of General Surgery
- Sahlgrenska University Hospital (Sahlgrenska Universitetssjukhuset)
  A university hospital (ca. 2400 beds)
  Department of Cardiothoracic Surgery

Blekinge Hospital, classed as a medium sized hospital in Sweden, is where we studied a general surgery department, whereas the study at Sahlgrenska University Hospital concerned specialised surgery that is mostly located at university hospitals.

Surgery Planning

The main problem is to map the list of planned patients into an operating schedule that meets both patient priority and the necessary available resources for a particular operation, whilst at the same time considering how the total care time is utilised, presuming that the objective is to operate on as many patients as possible. In this context, it is important to state that disturbances due to acute operations are not included explicitly in this study. By acute operations we mean patients that need an immediate operation. Based on medical assessments, patients are assigned into three priority groups;

1. Double priority (Operation needed within 2-4 weeks. Local differences)
2. Single priority
3. No priority

The surgical suite is resourced by a fixed number of operating rooms. In order to perform an operation there are some general requirements;

- One or more *surgeons*. The surgical skill must match the type of operation that is planned.
- One *operating room*
- One *operating team* consisting of (local differences):
    - one nurse anaesthetist,
    - one or two operating room nurses and
    - one or two assistant nurses. In our model, we include the *operating team* as available during the hours each operating room is open.
- One available anaesthetist (Not considered in this model)
- Special equipment, which might mean that some operations have to be performed in certain operating rooms. (Not considered in this model.)
- Post operative care. There must be beds available in the ICU (Intensive Care Unit) or the ward.

In order to handle priority group 1 and bearing in mind that we are in reality dealing with a rolling horizon, the final operation schedules are usually made one week at a time. In addition there is typically a waiting list of patients who can be operated on if time permits.

Department of General Surgery

In general surgery the surgeons are divided into teams by the hospital according to specialty. Table 1 shows an example of how the surgeons are grouped into surgical teams and how diagnosis/patients can be spread among them. Patient 1 represents a diagnosis which can be operated on by all surgical teams and all surgeons, a so-called "standard operation". The diagnosis represented by patient 2, requires a surgical operation handled by the surgeons of surgical team 1. Patient 5 has a diagnosis that requires skills, that only surgeon 1 and 2 (working in the same surgical team) have. Finally, patient 6 has a diagnosis that only can be treated by surgeons 6 and 9, but in contrast to the case of patient 5, the surgeons are working in different teams.

		Patient with example diagnosis					
		Patient 1	Patient 2	Patient 3	Patient 4	Patient 5	Patient 6
Surgeon team 1	Surgeon 1	ok	ok			ok	
	Surgeon 2	ok	ok			ok	
	Surgeon 3	ok	ok				
	Surgeon 4	ok	ok				
	Surgeon 5	ok	ok				
Surgeon team 2	Surgeon 6	ok		ok			ok
	Surgeon 7	ok		ok			
	Surgeon 8	ok		ok			
Surgeon team 3	Surgeon 9	ok			ok		ok
	Surgeon 10	ok			ok		
	Surgeon 11	ok			ok		

Table 1: Surgeon team and patient diagnosis.

Currently, each surgical team has a fixed number of operating rooms per week at its disposal. This is, as far as we are aware a common planning strategy at many medium sized Swedish hospitals. It is interesting to note how the surgical teams are allocated the same number of operating rooms every week with no consideration being given as to how the current patient load is distributed among the surgical teams or the availability of surgeons. In addition, surgeons from general surgery departments are also scheduled other duties in wards or consulting-rooms, which complicate the scheduling of the surgeons. In order to take other surgeon duties into account, surgeons cannot operate on two subsequent days in order to allow them to see patients, operated on the day before, in the ward and to see new patients in the consulting-room.

Department of Cardiothoracic Surgery

Due to the "complexity" of cardiothoracic surgery, operating teams are specialized and only work within cardiothoracic surgery. Unlike the surgeons in general surgery, the cardiothoracic surgeons are not occupied as much with other duties. This is explained by the fact that there are usually other physicians (generally cardiologists), rather than the cardiothoracic surgeon, that are responsible for the medical care and rehabilitation. This enhances the availability of the cardiothoracic surgeons considerably when compared to the general surgeons, and introduces greater flexibility into the surgeon scheduling. The internal divisions of surgical teams are structured in the same way as in general surgery. Also the principles of how the division of surgeons is related to different diagnosis, illustrated in Table 1, are also utilized in cardiothoracic surgery. Central to cardiothoracic surgery is the availability of ICUs (Intensive Care Units), which must have sufficient resources of beds and staff to handle the post operative care needs, i.e. patients transferred from the operating room suite. This in turn means that the wards must be able to accept patients transferred from the ICU to prevent restriction of patient through-put.

Summary

The operating room scheduling involves the same basic principles at both hospitals. Temporary schedules are constructed several weeks in advance, while the final schedules are constructed on the Thursday or Friday of the week before. The scheduling is performed manually (without any advanced software support). When comparing the two strategies for operating room scheduling, the surgeon schedules for general surgery deal with more aspects than in the cardiothoracic surgery scheduling. By aspects we mean that the individual needs of the surgeons in general surgery have to be considered, for example, not operating on two subsequent days. The static operating room allocation used in the case of general surgery is not used in the case of cardiothoracic surgery. The surgeons in the cardiothoracic surgery mostly operate during working hours and are not scheduled as much for other duties, as are the surgeons in the general surgery. An interesting observation related to operating room planning that is very valid for both of the case studies, is the importance of available resources in the ICU and later at ward level to prevent the restriction of patient through-put in the operating room suite, with the associated cancellation of operations. The latter could be important in considering the refinement of the processes in a logistics chain, to ensure sufficient post-operative care is available to meet the full operating room schedule. This parallel is sharper in the case study for Sahlgrenska University Hospital.

Problem Formulation

Based on the two case studies, we have identified the key resources that we find the most relevant when optimizing operating room planning. Also we have identified some rules related to the identified resources that have to be considered in order to meet additional non-key requirements.

Optimization Model

Indices and Sets:

j	Index for patient set J,
k	Index for surgeon team set K,
l	Index for surgeon set L,
m	Index for operating room set M,
t	Index for time slot set T.

Parameters:

a_{mt}	Available time in surgeon room m in time slot t (when time for potential acute operations has been excluded).
b_k	Parameter of how many operating rooms a surgeon team is allowed to allocate in the same time slot.
c_j	Cost of not operating on patient j.
d_{jl}	1 if surgeon l is qualified to operate on patient j, 0 otherwise.
e_j	Estimated operating time for patient j.
f_{lt}	1 if surgeon l is available on time slot t, 0 otherwise.
g_{kl}	1 if surgeon l is included in surgeon team k, 0 otherwise.
q_j	Estimated days of post operative care for patient j.

Variables:

o_{kmt}	1 if surgeon team k is allocated to operating room m in time slot t, 0 otherwise (binary).
p_t	Number of patients transferred from ICU (Intensive Care Unit) in time slot t (integer).

s_j 1 if patient j is not operated on, 0 otherwise (binary).

w_t Available beds at ICU (Intensive Care Unit) or the ward in time slot t (integer).

x_{lmt} 1 if surgeon l is assigned to operating room m in time slot t, 0 otherwise (binary).

y_{jmt} 1 if patient j is assigned operating room m in time slot t, 0 otherwise (binary).

z_{kt} 1 if surgeon team is operating in time slot t, 0 otherwise (binary).

Objective

The objective of operating room planning is to operate on as many patients as possible taking consideration of patient priority and current resources available. In our model we have chosen the objective function of minimizing the cost of not operating on a patient, which, as far as we know, is not the usual way of modelling. The cost parameter is meant to represent an estimation composed of a combination of patient suffering (related to; diagnosis, time waited for operation and medical priority) and public cost (e.g. cost of sickness benefits etc.) and is individually tailored to the patient. Alternatively, by having the cost parameter only representing one of the included issues, one cost perspective could be viewed.

Also, at this time, we make no consideration of the long term perspective and a rolling horizon, i.e. we are only considering one week of planning.

Minimize z = $\sum_{j \in J} c_j s_j$

Subject to:

Patients can only be operated on once. Also if operation of patient j does not occur for the current week, a cost is paid.

$$\sum_{m \in M} \sum_{t \in T} y_{jmt} = 1 - s_j \qquad \forall j, s_j \in \{0,1\} \qquad (1)$$

Capacity limit, the patients are associated with an estimated operating time according to the type and severity of the operation.

$$\sum_{j \in J} y_{jmt} e_j \leq a_{mt} \qquad \forall m,t \qquad (2)$$

Connection between patient diagnosis/operation types, surgeon available:

$$y_{jmt} \leq \sum_{l \in L} d_{jl} \cdot f_{lt} \cdot x_{lmt} \qquad \forall j,m,t \qquad (3)$$

Aspects of Surgeon Team and Operating Room Allocation

As described above, the surgical teams, in general surgery, are fixed to certain operating rooms and days (time slots) and within these assigned time slots and operating rooms, the surgeons within the surgeon team are distributed. As we are also studying the case of cardiothoracic surgery, where a dynamic/flexible operating room allocation is employed, we provide alternatives within the optimization model for analysis. To enable a flexible operating room allocation, we set up limitations for, on the one hand, the surgical teams and, on the other hand, the surgeons, to operate on two subsequent days by replacing the fixed operating room allocation with either constraint (4a) or (4b). By this, we keep the restrictions of having the surgeons free from operative duties the day after having operated in order for them to be accessible for post operative patients as required in general surgery.

$$z_{k(t-1)} + z_{kt} \leq 1 \qquad \forall k,t, \quad t \geq 2 \qquad (4a)$$

$$\sum_{m \in M} (x_{lm(t-1)} + x_{lmt}) \leq 1. \qquad \forall k,t, \quad t \geq 2 \qquad (4b)$$

Constraints (5) are related to the choice of implicating constraint (4a) and they limit how many operating rooms a surgeon team is allowed to occupy at the same time.

$$\sum_{m \in M} o_{kmt} \leq b_k * z_{kt} \qquad \forall k,t \qquad (5)$$

Constraint ensuring that only one surgeon team is allowed to occupy an operating room in a time slot.

$$\sum_{k\in K} o_{kmt} \leq 1 \qquad \forall m,t \qquad (6)$$

Restricting surgeons to only one surgeon team per operating room and time slot:

$$x_{lmt} \leq \sum_{k\in K:g_{kl}=1} o_{kmt} \qquad \forall l,m,t \qquad (7)$$

Restrict the surgeon to one operating room per time slot.

$$\sum_{m\in M} x_{lmt} \leq 1, \qquad \forall l,t \qquad (8)$$

Model Extension

One of the main causes of disturbances to operating room planning, besides the unpredictable occurrence of acute operations, is the restriction of patient through-put in either the ICU or the ward. The possibility of including estimated expected post operative care into the optimization model for the operating room planning is therefore interesting. In the following experiment description, constraint (9) and (10) are not included.

Available ICU/Ward-beds

$$w_{(t-1)} + p_t - \sum_{j\in J}\sum_{m\in M} y_{jmt} = w_t \qquad \forall t \qquad (9)$$

Patients transferred from ICU/Ward in time slot *t*.

$$p_t = \sum_{j\in J}\sum_{m\in M} y_{jm(t-q_j)} \qquad \forall t \qquad (10)$$

We assume that $y_{jm(t-q_j)}$ is a given parameter for $t - q_j \leq 0$ (i.e. given by earlier operations).

Operating Room Scheduling

Based on the two conducted case studies, we analyze the potential of optimization modelling for operation room planning to support the allocation of key-resources. However, we focus on the general surgery and have conducted an experiment using the provided model to help analyze static versus dynamic operating room allocation. As pointed out above, there is a difference in direction between how the allocation of operating rooms and scheduling of surgeons is performed in the two hospitals. The experiment set-up and results are further described below.

Experiment

The main outline of the conducted experiment of scenarios is organized according to:

1. Analysis concerning static allocation of operating rooms compared to dynamic/flexible allocation of the same.
2. Extended opening hours of the operating rooms.

When utilizing static operating room allocation, the surgical teams allocate the same number of rooms on the same day (time slot) every week. While in the dynamic/flexible approach, employed in cardiothoracic surgery, the surgeons (not the surgical teams) allocate the rooms and time slots according to patient load. This fact gave us the idea of allowing for the optimization model to incorporate a choice of the level of flexibility, in order to simulate the dynamic/flexible approach in general surgery as well. We have modelled (see description of optimization model) so that we are free to choose either a *static surgical team allocation*, as is utilized in Blekinge Hospital, *flexible surgical team allocation* by applying constraint (4a), or the alternative of no team allocation; *flexible surgical allocation* by applying constraint (4b), utilized in Sahlgrenska University Hospital. Constraint (9) and (10) are, as earlier mentioned, not applied in this experiment principally because further investigations into how the estimated post operative care can be related to the clinic (patient group) are needed.

We use opening hours for each operating room, denoted default time, according to Table 2 below (parameter a_{mt}). Then we extend opening hours to 450 minutes on Mondays and Tuesdays in Room 1 to illustrate the effects of potential shifts, over-time or flex-time for the staff (operating room team). No considerations as to the availability of staff and beds are taken. On Fridays, the time slots are smaller due to staff constraints during week-ends.

	Monday	Tuesday	Wednesday	Thursday	Friday
Room 1	350min	350min	300min	300min	200min
Room 2	300min	300min	300min	300min	200min
Room 3	300min	300min	300min	300min	200min

Table 2: Default opening hours

In order to conduct as realistic experiment scenarios as possible, we use input data based on statistics from Blekinge Hospital. The scenarios represent one week of operating room planning. The numbers and types of operations that apply for operating room admission in the scenarios are in ratio to operations performed during the last year at Blekinge Hospital. We have chosen the number of 40 patients at each scenario to represent the patient queue. No consideration of seasonal variations is taken, if indeed there are any. The estimated length of time for each operation, e_j, is based on mean values from real data plus 45 minutes of set-up time (also mean value). Also the number of surgeons and surgical teams correspond to the approximate number of available surgeons during one week at Blekinge Hospital. In this experiment we use five surgical teams with three surgeons each. The individual schedules for the surgeons, i.e. what day to operate or to have a day-off and so on, are pre-determined, which together with the fixed allocation of operating room constitute the basis of the current operating room scheduling at Blekinge Hospital and is here denoted; *default case*. To enable flexible room allocation we have allowed the individual surgeon schedules to remain in a preliminary state. This implies that the surgeons are not scheduled for surgery or other duties until the final surgery schedule is set. We scheduled two of the three surgeons to be always available for surgery. The number of operating rooms considered is three, as in Blekinge Hospital, and the time slots representing opening hours per day and room are calculated as four full-time days, Monday-Thursday, and one half day on Friday. The opening hours for each room are in reality dependent on the availability of the operating team (not to confuse with the surgeons). The cost parameters c_j are not based on real cases but are here randomized to represent patient diagnosis and patient waiting time. Our intention is to further develop this value into realistic patient cost parameters.

As described earlier, the surgeons in general surgery are often responsible for the medical care of the patients. In several cases this implies the responsible surgeon also needs to be the operating surgeon, i.e. a combination of patient-surgeon that is fixed. This could also be the case when some operations require special surgical qualifications. Patients with double priority often have contracted diseases with carcinogenic or other problematic characteristics which involve additional needs. In reality, this means, the surgeon together with the patient needs to decide what day the operation will be performed. To reflect these needs and come as close to reality as possible, we have constructed six fixed operations per week distributed among all surgeon teams, for the scenarios which are not default. This entails increased validity for the suggested dynamic approach, and allows us to compare our proposal to the existing system (here; default scenarios) of operating room planning. We use 40 patients for admission to the operating room planning as our default. Each patient is assigned a random cost between 1–70, except for 6 patients, who are assigned the value of 100 to represent the patients that need to be scheduled within this time frame. This cost is modelled as a penalty to the objective function when the patient is not operated on and represents a mix of medical priority, time waited and economic aspects. Except for the following described differences, the scenarios use the same system settings for optimization.

Results

We have used 4 sets of data input to represent the patient associated cost parameter when optimizing. In Table 3 we can first see the minimized costs from the default scenarios, i.e. *Fixed Allocation.* The total computed cost results indicate that using a more dynamic/flexible operating room planning model is more cost-efficient than the default case. The flexible approach implies a general increase in efficiency, i.e. increased patient throughput.

	Default opening hours				Extenden opening hours			
	1	2	3	4	1	2	3	4
Fixed Allocation	157	173	191	117	123	142	153	153
Flexible Surgeon Team Allocation	132	149	150	100	72	72*	83*	42*
Flexible Surgeon Allocation	121	149*	150*	84*	54	72*	83*	42*

Table 3: Minimized cost according to allocation strategy and opening hours.

In the table the asterisk denotes some better solutions might exist since the gap in the branch and bound approach (Cplex 8.1) could not be closed within the time limit of 1000 seconds of CPU-time.

Conclusion and Future work

One of the main objectives for this research has been to provide optimization modelling to support analyzing operating room planning. Although this experiment only shows preliminary results, we have demonstrated how optimization modelling can be used for analyzing strategies and different settings with respect to resource allocation. The experiment was conducted in collaboration with two Swedish hospitals, which to a certain extent, affected the problem formulation. The conclusions have to be interpreted with this in mind, e.g. the amount of flexibility in allocation depends on the organizational structures and might differ between different countries. The results from the experiment were demonstrated to Blekinge Hospital provoking interested reactions. However, we are aware that the model is limited and our intention is to expand the model in different directions. For example, further research with experiments, with constraints (9) and (10), considering the logistics to a greater extent, are already in progress and the results will soon be reported. We will further study the impact of a more representative cost parameter on the results. Also, the omitted aspect of operating room planning under uncertainty, i.e. acute operations and other events, is very relevant and will be considered for future work. Since the size of the operating room allocation problem could be extensive, we are looking into additional optimization techniques, such as decomposition, for better and faster results.

Acknowledgements

Funding for this research has been provided by County Council of Blekinge. The Department of General Surgery at Blekinge Hospital and Department of Cardiothoracic Surgery and Department of Cardiothoracic Anaesthesia and Intensive Care at Sahlgrenska University Hospital have generously made data available to us.

References

H. Beaulieu, J. A. Ferland, B. Gendron, P. Michelon, (2000). A mathematical programming approach for scheduling physicians in the emergency room, *Health Care Management Science* 3 193-200.

V. De Angelis, G. Felici, P. Impelluso (2003). Integrating Simulation and optimisation in health care centre management, *European Journal of Operational Research* 150 101-114.

B. Hollingsworth (2003). Non-Parametric and Parametric Applications Measuring Efficiency in Health Care, *Health Care Management Science* 6, 203–208.

J.B. Jun, S.H. Jacobson, J.R: Swisher (1999). Application of discrete-event simulation in health care clinics: A survey, *Journal of the Operational Research Society* 50, 109–123.

P.C. Kua, R.A. Schroeder, S. Mahaffey, R.R. Bollinger (2003). Optimisation of Operating Room Allocation Using Linear programming Techniques, *American College of Surgeons.*

M. Lagergren (1998). What is the role of models to management and research in the health services? A view from Europe, *European Journal of Operational Research* 105, 257–266.

J.C. Lowery (1992). Simulation of a hospital's surgical suite and critical care area, *Proceeding of the 1992 Winter Simulation Conference.*

S.N. Ogulata, R. Erol (2003). A Hierarchical Multiple Criteria Mathematical Programming Approach for Scheduling General Surgery Operations in Large Hospitals, *Journal of Medical Systems*, Vol. 27, No. 3.

L.A. Riley (1999). Applied Simulation as a Decision Support System Tool: The Design of a New Internal Medicine Facility, *Proceeding of the 32nd Hawaii International Conference on System Sciences.*

NSPLIB – A NURSE SCHEDULING PROBLEM LIBRARY: A TOOL TO EVALUATE (META-) HEURISTIC PROCEDURES

MARIO VANHOUCKE[1, 2] and BROOS MAENHOUT[1]
[1]*Faculty of Economics and Business Administration, Ghent University, Gent, Belgium.*
[2]*Operations & Technology Management Centre, Vlerick Leuven Gent Management School, Gent, Belgium.*

Abstract

In this paper, we propose a large set of problem instances based on different complexity indicators for the well-known nurse scheduling problem (NSP). The NSP assigns nurses to shifts per day taking both hard and soft constraints into account. The objective is to maximize the overall quality of the nurse roster by taking nurses' preferences and the penalty costs of violating the soft constraints into account. The problem is known to be NP-hard.

Due to its complexity and relevance in practice, many researchers have developed (meta-)heuristic procedures to solve a NSP instance heuristically in an acceptable time limit. However, these solution procedures are often very case-specific towards one hospital and hence, cannot be compared with each other. Moreover, lack of data and the many interpretations of how to evaluate solution procedures, have contributed to the never-ending amount of newly developed procedures without any published effort to benchmark them.

The contribution of this paper is threefold. First, we propose a large set of benchmark instances for the nurse scheduling problem in order to facilitate the evaluation of existing and future research techniques. Secondly, we propose a computer platform independent stop criterion to evaluate and compare meta-heuristic procedures for the NSP. Finally, we propose a newly developed website where the benchmark instances can be downloaded and where the best known solutions can be uploaded.

Keywords: manpower planning; hospitals; libraries.

Introduction

The nurse scheduling problem (NSP) is a well-known combinatorial optimization problem in literature and has attracted numerous researchers to develop exact and (meta-)heuristic procedures. The NSP involves the construction of duty rosters for nursing staff and assigns the nurses to shifts per day taking both hard and soft constraints into account. The objective is to maximize the overall quality of the nurse roster by taking nurses' preferences and penalty costs of violating the soft constraints into account. The problem is known to be NP-hard (Osogami and Imai, 2000).

Due to its complexity and relevance in practice, the operations research literature has been overwhelmed by different procedures to solve the problem. The complexity has resulted in the development of several (meta-)heuristic procedures, able to solve a NSP instance heuristically in an acceptable time limit. The practical relevance has resulted in a never-ending amount of different NSP versions, taking practical, case-specific constraints into account. Despite the numerous procedures for the NSP, no state-of-the-art results have been presented in literature. The main reason is that comparison between procedures is very difficult, since problem descriptions and models vary drastically and depend on the need of the particular hospital. Due to the huge variety of hard and soft constraints, and the several objective function possibilities, the nurse scheduling problem has a multitude of representations which have resulted in the wide variety of different solution procedures. The comparison between procedures is further hindered by the lack of benchmark problem instances and the unavailability of source code of the different procedures. Moreover, there is no general agreement of how to evaluate and compare procedures in terms of solution comparison, stop criterion, etcetera. Consequently, a fair comparison between procedures seems to be an impossible idea, which undoubtedly limits the efficient development of future algorithms. Note that both Cheang *et al* (2003) and Burke *et al* (2004) express the need for a benchmark database to facilitate comparisons of the various algorithms and to motivate future researchers to develop better solution procedures for the NSP.

In this paper, we approach the identified need of benchmarking by presenting a large set of problem instances for the NSP. The outline of this paper is as follows. In the next section, we highlight the necessary conditions under which a benchmark dataset can be used by the operations research community. After this we present the details of our nurse scheduling problem library NSPLIB. In this section, the complexity of the NSP is discussed, followed by the settings and details of our problem instances. Moreover, we propose a computer platform independent stop criterion so that our problem instances allow a fair evaluation and comparison between different meta-heuristic procedures for the NSP. Finally we make conclusions and highlights avenues for future research.

Benchmark Libraries in the OR Community

One of the most important applications of a benchmark library is to provide a common set of problem instances on which different researchers can easily benchmark their algorithms. In doing so, the library provides a competitive environment among researchers, which stimulates progress in the development of algorithms and contributes to the gaining of insights into the performance of these algorithms. Hooker (1995) calls our attention to potential pitfalls of *competitive testing*, and argues that benchmark instances should stimulate scientific testing rather than only create a competitive environment where algorithms are evaluated purely on computational performance. *Scientific testing*, on the contrary, still recognizes that fast and efficient codes should be written, but puts more focus on the development of algorithms to yield insight into their performance. As a consequence, we believe a benchmark dataset should satisfy the following conditions (these four conditions, among others, have also been mentioned by Gent and Walsh (1999)):

- **Diversity:** the problem set should contain instances that are as diverse as possible. Hoos and Stützle (2000) argue that benchmark sets should contain a large variety of different types of problem instances, so that they can be used for evaluating different types of algorithms in an as unbiased way as possible. Elmaghraby and Herroelen (1980) argue that diversity of problem instances (which are, in their case, project networks) is indispensable in order to span *the full range of complexity*. Hooker (1995) argues that a controlled experimentation begins with a selection of *n* factors that could affect performance, in order to have a diverse set of problem instances which are homogeneous with respect to characteristics that are likely to affect performance.

- **Realism:** Since the ultimate goal of research is to develop algorithms for solving real-life problems, a benchmark library should reflect real-life problems.

- **Size:** Ideally, a benchmark dataset should be as large as possible. However, testing algorithms often involves a trade-off between problem instance size and solvability (due to the heavy burden of required CPU time). Hence, we have chosen to vary the size of our problem instances from rather small to sufficiently large instances, in order to allow the researcher to draw general conclusions on his/her selected test scenarios.

- **Extensibility:** The problem set should be easy to extend to other features. A dynamic benchmark dataset is preferable to a static one (Hooker, 1994 and 1995), since the latter assumes that a problem description (like the NSP) never changes. However, new features can show up as progress is made, coming from new research avenues or as a result of practical needs. The dataset should allow the incorporation of these new features in an easy and comprehensible way.

In the remainder of this paper, we will propose different test scenarios that fulfil the requirements above. In the next section, we propose both a diverse set and a realistic set with varying size, that are easily extendable to other features.

NSPLIB: the Nurse Scheduling Problem LIBrary

Complexity of the NSP

In the basic nurse scheduling problem, a set of nurses needs to be assigned to one of a number of possible shifts in order to meet the minimal coverage constraints and other case-specific constraints, and to maximize the quality of assigned working shifts. According to Warner (1976), quantifying *preferences* in the objective function maintains fairness in scheduling nurses over the scheduling horizon. Hence, the quality of a schedule is a subjective judgment of the nurses, depending on how well the assigned schedule conforms to his/her desire to be off or on duty, and to other schedule properties such as work stretch, rotation patterns, etc.... The *coverage constraints* determine the required nurses per shift and per day, and are inherent to each NSP instance. However, many other constraints are very *case-specific*, and are determined by personal time requirements, specific workplace conditions, national legislation, etc.... Consequently, each data instance should contain information on the following three classes in order to be useful for a NSP procedure:

- Preference matrix: the preference or aversion of nurses to work on a shift/day
- Coverage constraints: the required number of nurses
- Other (often very case-specific) constraints

The first two classes describe a two-dimensional nurse/day roster matrix, which is inherent to any NSP instance. To that purpose, Vanhoucke and Maenhout (2005) have proposed nine complexity indicators in order to describe a NSP roster matrix and have presented a generator to construct NSP instances based on these nine input parameters. These indicators can be used to generate data under a controlled design to measure the size and the structure of the preference matrix and the corresponding coverage requirements. Hence, these indicators do not describe the case-specific constraints.

The size of the NSP instance under study depends on the size of the duty roster matrix. Hence, the size of a NSP instance can be measured by three indicators, as follows:

- Number of nurses
- Number of days
- Number of shifts

Three indicators measure the structure of the preference matrix, as follows:

- Nurse-preference distribution (*NPD*): Distribution of preferences among the nurses
- Shift-preference distribution (*SPD*): Distribution of nurse's preferences among shifts
- Day-preference distribution (*DPD*): Distribution of nurse's preferences among days

The last three indicators measure the way the coverage requirements are distributed among shifts and days, as follows:

- Total coverage constrainedness (*TCC*): Total number of nurses required
- Shift-coverage distribution (*SCD*): Distribution of coverage requirements among shifts
- Day-coverage distribution (*DCD*): Distribution of coverage requirements among days

For more information about the calculation and interpretation of these indicators, we refer to the working paper of Vanhoucke and Maenhout (2005).

The library

The development of exact and (meta-)heuristic procedures for the nurse scheduling problem fuelled the need for benchmark instances (Cheang *et al*, 2003; Burke *et al*, 2004). Rather than testing a newly developed algorithm on one real-life NSP instance (which has high practical relevance), one can additionally test its procedure on a dataset and report results. In doing so, a benchmark dataset will be shared among the research community that will facilitate the systematic evaluation and comparison of the performance of the different procedures.

In this subsection, we discuss three aspects of the nurse scheduling problem library (NSPLIB): 1) two sets of problem instances are presented that are accessible by the research community 2) we show how to incorporate classes of case-specific constraints, and finally 3) we propose a way to test and evaluate procedures that is independent of the computer platform used.

Benchmark instances

The benchmark instances have been grouped in two different sets; a diverse and a realistic set, each containing four and two sub-sets, respectively. Each sub-set is characterized by systematically varied levels of the complexity indicators. Table 1 contains the values for the different levels of the nine complexity indicators.

The diverse set has been constructed using three levels for the last six complexity indicators, in order to guarantee that the problem instances span the full range of complexity. Although we varied the number of nurses from 25 to 100, we kept the planning horizon rather small (7 days). Using ten instances per setting, each sub-set contains 7,290 problem instances. This large number of benchmark instances has been constructed to test meta-heuristic procedures in an acceptable time (small planning horizon) but prevents the over-fitting of the tested procedures due to the large number of instances, i.e. 29,160 in total.

The realistic set has been constructed with only two levels for the last six complexity indicators, but within a planning horizon of 28 days. Since many procedures in literature have been developed for and tested within a planning horizon of 28 days, we believe that this set is crucial and indispensable to compare and benchmark the current state-of-the-art NSP procedures. Using ten instances per setting, each sub-set contains 960 instances.

Diverse Set			Realistic Set	
Problem size			**Problem size**	
N	25, 50, 75 or 100		*N*	30 or 60
S	4 (including the free shift)		*S*	4 (including the free shift)
D	7		*D*	28
Preference distribution			**Preference distribution**	
NPD	0.25, 0.50 or 0.75		*NPD*	0.3 or 0.7
SPD	0.25, 0.50 or 0.75		*SPD*	0.3 or 0.7
DPD	0.25, 0.50 or 0.75		*DPD*	0.3 or 0.7
Coverage constraints			**Coverage constraints**	
TCC	0.20, 0.35 or 0.50		*TCC*	0.20, 0.35 or 0.50
DCD	0.25, 0.50 or 0.75		*DCD*	0.3 or 0.7
SCD	0.25, 0.50 or 0.75		*SCD*	0.3 or 0.7

Table 1: Test settings for our benchmark instances

The problem instances can be downloaded from www.projectmanagement.ugent.be/nsp.php under the names N25.zip, N50.zip, N75.zip and N100.zip (diverse set) and N30.zip and N60.zip (realistic set). Note that, to the best of our knowledge, only two other NSP datasets are available on the internet by Burke *et al* (2004) and Musliu *et al* (2004). The first set contains real world data and can be downloaded from:
http://ingenieur.kahosl.be/vakgroep/it/nurse/archive.htm.

The second set contains a randomly generated dataset for the rotating workforce scheduling problem, and can be downloaded from:
http://www.dbai.tuwien.ac.at/staff/musliu/benchmarks/.

Case-specific constraint

In order to solve nurse scheduling problem instances, many more case-specific constraints can be incorporated into the problem description. Cheang *et al* (2003) identified nine constraint types as appearing frequently in the literature. We have extended the NSPLIB with case constraint files (CC-files) which incorporate a subset of these nine constraints with specific input parameters. Each file can be used as an additional set of constraints for the proposed benchmark instances. The daily coverage requirements are an inherent characteristic of all personnel rostering problems, and are incorporated in three of the aforementioned indicators. Moreover, two other constraints (i.e. minimal free time between working shifts and the fact that nurses can only take one assignment per day) occur almost always in literature (Cheang *et al*, 2003). These three constraints are supposed to characterize all nurse scheduling problems and are applicable in all cases (Koop, 1998). Furthermore, some constraints could be distinguished which frequently typify nurse scheduling practices but are not always incorporated. To that purpose, we have classified the case-specific constraints as follows:

- Number of assignments, i.e. non-free shift assignments per scheduling period
- Number of assignments per shift, i.e. identical shift assignments per scheduling period
- Consecutive working shifts, i.e. non-free shift assignments per scheduling period
- Consecutive same working shifts, i.e. identical shift assignments per scheduling period.

Not only is the presence of these constraints an indicator of the restrictiveness of nurse scheduling instances, but also the parameter setting for each constraint contributes to the restrictiveness of this problem. For each of these cases, some settings are defined based on specific cases described in literature, or on interesting settings not mentioned in literature. An overview of the sixteen cases containing a mix of case-specific constraints and their corresponding settings are given in table 2. The minimal value and maximal value is given between brackets and can be different for each shift. As an example, CC-file 7 requires a minimal 2 consecutive assignments and maximum 3 for the three working shifts (no requirements are given for the free shift), denoted by [2,3], [2,3], [2,3]. The detailed format of each CC-file and the specific settings of each constraint are given on the website and are outside the scope of this paper.

Evaluation of (meta-)heuristic pProcedures for the NSP

We would like to request that researchers who test their specific procedure on the proposed test design contribute to the development of best-known solutions which can be used for future comparison purposes. However, in order to make a fair comparison between procedures, one needs a clear and easily applicable stop criterion that is independent of the computer platform and coding skills. In recent research papers, many algorithms have been developed, but little effort has been put into standardizing the test approach. Table 3 displays the different meta-heuristic approaches for the NSP. For each method, we have listed the stop criterion, the required computational time, the test instances used as well as information about the problem size and the computer platform.

	Number of assignments	Number of assignments per shift	Consecutive working shifts	Consecutive same working shifts
	Diverse Set: N25, N50, N75 and N100 instances			
Case 1	[5,5]	x	x	x
Case 2	[4,6]	x	x	x
Case 3	[5,5]	[1,3], [1,3], [1,2]	x	x
Case 4	[4,5]	[0,5], [0,5], [0,4]	x	x
Case 5	[5,5]	x	[2,5]	x
Case 6	[4,6]	x	[1,5]	x
Case 7	[5,5]	[0,5], [0,5], [0,3]	[2,5]	[2,3], [2,3], [2,3]
Case 8	[2,6]	[0,6], [0,6], [0,3]	[2,4]	[1,4], [1,4], [2,4]
	Realistic Set: N30 and N60 instances			
Case 9	[20,20]	x	[1,7]	[1,7], [1,7], [1,7]
Case 10	[16,24]	x	[1,7]	[1,7], [1,7], [1,7]
Case 11	[20,20]	[4,12], [4,12], [4,8]	[1,7]	[1,7], [1,7], [1,7]
Case 12	[16,20]	[0,20], [0,20], [0,16]	[1,7]	[1,7], [1,7], [1,7]
Case 13	[20,20]	x	[2,5]	[1,7], [1,7], [1,7]
Case 14	[16,24]	x	[1,5]	[1,7], [1,7], [1,7]
Case 15	[20,20]	[0,20], [0,20], [0,12]	[2,5]	[2,3], [2,3], [2,3]
Case 16	[16,24]	[0,24], [0,24], [0,12]	[2,4]	[1,4], [1,4], [2,4]

Table 2: The case-specific constraints per CC with [min, max] settings

We have copied the original terminology of the papers (e.g. the term 'generations' is often used for genetic algorithms while 'moves' are used in tabu search algorithms), without explaining its details. Inoue *et al* (2003), for example, clearly describe a generation as a combination of crossover operations, heuristics and mutation. In other papers, the stop criteria are somewhat vaguer. Therefore, we would like to propose a test design based on a clear and easily applicable stop criterion. This test design is very much inspired by the critical remarks on the use of the test design written by Kolisch and Hartmann (2005). These authors use the number of created schedules as a stop criterion for meta-heuristic procedures in project scheduling. In doing so, the criterion is independent of the computer platform, and hence, allows a fair comparison between procedures. Inspired by these ideas and given the competitive environment in which researchers develop project scheduling procedures which outperform previous ones, we would like to call for the use of the *number of visited solutions* as the one and only stop criterion for testing meta-heuristic nurse scheduling problem procedures. Although this stop criterion can also be criticised as a potential pitfall, we would like to minimise the confusion about when a solution can be considered as visited (and hence, accounts for an extra solution closer to the stop criterion). Therefore, from the moment the algorithm evaluates the objective function of the new solution, an additional solution has

Reference[1]	Stop Criterion	CPU	Dataset	Problem Size[2] N, D, S	Computer Platform
Aickelin and Dowsland, 2000 (GA)	No improvement during 30 generations	14.9s	52 real problem instances	30, 7, 4	Pentium II 200
Aickelin and Dowsland, 2004 (GA)	100 generations	9.3s	52 real problem instances	30, 7, 4	Pentium II
Bellanti et al., 2004 (TS)	No improvement during 50 neighbourhoods scans	2s up to 6h47m36s	4 real problem instances / 12 random instances	20 to 60, 30, 4	Pentium IV, 2Ghz
Burke et al., 1998 (TS)	No improvement during x moves	1m28s up to 28m8s	2 real problem instances	20, 28, 8	IBM Power PC RS6000
Burke et al., 2001 (TS/GA/MA)	No improvement during 2 generations	49s up to 2h31m51s	4 real problem instances	20, 28, 8	IBM Power PC RS6000
Burke et al., 2005 (TS)	No improvement during x moves	3m41s up to 23m19s	1 real problem instance	20, 28, 8	IBM Power PC RS6000
Cowling et al., 2003 (TS/GA)	6,000 iterations	44s up to 60s	52 real problem instances	30, 7, 4	Pentium II, 1000 Mhz
Dias et al., 2003 (GA/TS)	×	ca 120s	12 real problem instances	30, 30, 4	AMD Athlon 600
Dowsland, 1998 (TS)	No improvement during 1,000 moves	60s up to 150s	real problem instances	Up to 30, 7 to 42, 4	Pentium 60
Inoue et al., 2003 (GA)	50 generations	×	1 real problem instance	20, 30, 6	×
Jan et al., 1999 (GA)	10,000; 100,000; 200,000 generations	49s	1 real problem instance	15, 30, 4	Pentium II 300
Kragelund and Mayoh, 1999 (SA)	No improvement during 1,000,000 moves or when penalty costs are zero	×	2 real problem instances	45, 30, 11	×
Li and Aickelin, 2003 (BA)	2,000 generations	10s up to 20s	52 real problem instances	30, 7, 4	Pentium 4
Louw et al., 2005 (TS)	No improvement >2% during x moves	1s up to 2h18m09s	1 real problem instance	22, 28, 3	667 Mhz

[1] We use the abbreviations GA for Genetic Algorithms, TS for Tabu Search algorithms, MA for Memetic Algorithms, BA for Bayesian optimization Algorithm and SA for Simulated Annealing algorithms.
[2] We use the notation N (number of nurses), D (number of days) and S (number of shifts) to describe the size of the roster matrix
× = no information available

Table 3: Test settings for the meta-heuristic procedures in literature been generated by the algorithm. This means that the mutation operations and local search algorithms contribute to the solution generation, only from the moment an evaluation takes place. Consecutive (small) changes to a nurse roster (e.g. a series of shift-pattern swaps between nurses) only results in a new solution when the effect on the objective function is known. This is completely in line with Kolisch and Hartmann (2005) who use a so-called schedule generation scheme, which basically involves the construction and evaluation of a project schedule, as a criterion to count an extra schedule.

The problem instances, the CC-files and the solutions can be downloaded from our website (www.projectmanagement.ugent.be/nsp.php). The solution files (in Excel) contain the individual solution of each problem instance for 1000 and 5000 schedules, as well as the best known solutions for the test instances (without a given stop criterion). For each problem instance, the total solution quality, the CPU-time, and the number of violations regarding the minimal coverage requirements have been reported. Up to now, the solutions are found by the procedure of Maenhout and Vanhoucke (2005) using their electromagnetism approach.

The website will be regularly extended, and we request researchers to report solutions of their procedures when they lead to an improvement (both within a stop criterion of 1,000 and 5,000 schedules and without a stop criterion). Moreover, newly tested (real-life) nurse scheduling problem instances and new case constraint files are welcome and contribute to the dynamic extension of our library. Finally, researchers who wish to generate other problem instances to adapt them to their specific settings can download the generator from the same web address. We are confident that we have created a competitive environment in which to stimulate researchers to outperform these known results by the development of more novel procedures.

Extension to Other Problem Types

In the previous section, we argued that the proposed library is open for extensions, and researchers are free to submit new (real-life) problem instances as well as new CC-files. We believe that our suggested datasets can be used for a wide variety of nurse scheduling problems, without altering the data, and might be good benchmark sets for a broader class of employee scheduling problems.

The incorporation of *grades among nurses* (i.e. employee skill level or employee status) implies day/shift coverage requirements for each grade, and assumes that coverage requirements for nurses with a particular grade can be fulfilled by nurses of a higher grade. The extension to different *types of nurses* (e.g. full-time nurses, part-time and nurses working four days a week, etc…) can also be easily incorporated in algorithms and does not require drastic changes to the data. Each nurse of our specific problem formulation needs to be assigned to one of the possible types or grades of nurses defined by the user. In both cases (types and grades) it might be necessary to define new CC-files to incorporate these extra features.

Employee scheduling problems arise in a variety of environments and hence, we believe that our problem instances can be used for other problem types outside hospitals. Glover and McMillan (1986) mention telephone operator scheduling, airline and hotel reservation personnel scheduling, bank personnel scheduling and many more applications in the service delivery settings. In their paper, they mention availability preferences that can be specified by each employee, which very much resemble the nurses' preferences included in the current paper. Hence, we believe that our current test scenarios are general and can be used in a wide variety of settings.

Conclusion

In this paper, we have presented a benchmark dataset for the well-known nurse scheduling problem. A nurse scheduling problem instance is characterised by a two-dimensional roster matrix, describing the nurses' preferences and related coverage requirements, as well as by other, often very case-specific, constraints. In our dataset, we have relied on nine indicators developed by Vanhoucke and Maenhout (2005) to control the two-dimensional roster matrix, and have presented so-called CC-files to incorporate case-specific constraints.

The proposed problem sets fulfil the four important criteria of a benchmark library, i.e. diversity, realism, varying size and extensibility.

First, our dataset guarantees *diversity* in order to span the full range of complexity (Elmaghraby and Herroelen, 1980). More precisely, we have generated a large diverse set, containing four sub-sets with six varying input parameters measuring the structure of the roster matrix. All problem instances have been constructed for a scheduling horizon of 7 days and 4 shifts.

Secondly, a second dataset serves as our *realistic* set, since we have generated problem instances that assume a four-week planning horizon and contains a much smaller amount of problem instances.

Thirdly, each sub-set of the previously mentioned dataset has varying *sizes*: the diverse sub-sets contain instances with 25, 50, 75 and 100 nurses while the realistic sub-sets contain instances with 30 and 60 nurses.

Last, our library fulfils the *extensibility* criterion since it is one of our aims to dynamically update the library with newly developed best known solutions and new real-life problem instances, as well as newly created case-specific constraint files.

All problem instances, as well as the obtained solutions and the CC-files can be downloaded from the website www.projectmanagement.ugent.be/nsp.php. The website will be regularly extended, and we strongly recommend researchers to report their new benchmark solutions on this site. Adding new CC-files as well as real-life problems are also part of our update intentions and contribute to the competitive research environment striving to solve different aspects of the nurse scheduling problem. We are confident that we have created a competitive environment to stimulate all researchers to make progress and benchmark their results with the best known published results.. We will spend time and effort in the maintenance of the website in order to facilitate the accessibility and encourage research progress.

References

Aickelin U and Dowsland K A (2000). Exploiting problem structure in a genetic algorithm approach to a nurse rostering problem. *J Scheduling* 3: 139–153.

Aickelin U and Dowsland K A (2004). An indirect Genetic Algorithm for a nurse-scheduling problem. *Comput Oper Res* 31: 761–778.

Bellanti F, Carello G, Della Croce F and Tadei R (2004). A greedy-based neighborhood search approach to a nurse rostering problem. *Eur J Oper Res* 153: 28–40.

Burke E K, De Causmaecker P and Vanden Berghe G (1998). A Hybrid Tabu Search Algorithm for the Nurse Rostering Problem. *Lect Notes Comput Sc* 1585: 187–194.

Burke E K, Cowling P, De Causmacker P and Vanden Berghe G (2001). A memetic approach to the nurse rostering problem. *Appl Intell* 3: 199–214.

Burke E K, De Causmaecker P, Vanden Berghe G and Van Landeghem H (2004). The state of the art of nurse rostering. *J Scheduling* 7: 441–499.

Burke E K, De Causmacker P, Petrovic S and Vanden Berghe G (2004). Variable neighbourhood search for Nurse Rostering Problems. In: Resende M G C and Pinho de Sousa J (eds). *Metaheuristics: Computer Decision-Making*, pp 153–172.

Cheang B, Li H, Lim A and Rodrigues B (2003). Nurse rostering problems – a bibliographic survey. *Eur J of Oper Res* 151: 447–460.

Cowling P, Kendall G and Soubeiga E (2002). Hyperheuristics: A Robust Optimisation Method Applied to Nurse Scheduling. *Lect Notes Comput Sc* 2439: 851–860.

Dias T M, Ferber D F, de Souza C C and Moura A V (2003). Constructing nurse schedules at large hospitals. *International Transactions in Operational Research* 10: 245–265

Dowsland K A (1998). Nurse scheduling with Tabu Search and strategic oscillation. *Eur J Oper Res* 106: 393–407.

Elmaghraby S E and Herroelen W S (1980). On the measurement of complexity in activity networks. *Eur J Oper Res* 5: 223–234.

Gent I P and Walsh T (2004). CSPLIB: A Benchmark Library for Constraints. *Lect Notes Comput Sc* 1713: 480–481.

Glover F and McMillan C (1986). The General Employee Scheduling Problem: An integration of MS and AI. *Comput Oper Res* 13: 563–573.

Hooker J N (1994). Needed: An empirical science of algorithms. *Oper Res* 42: 201–212.

Hooker J N (1995). Testing Heuristics – We Have It All Wrong. *J Heuristics* 1: 33–42.

Hoos H H and Stützle T (2000). SATLIB: An Online Resource for Research on SAT. In Gent I P, van Maaren H and Walsh T (eds). SAT20000: Highlights of Satisfiability Research in the year 2000: Frontiers in Artificial Intelligence and Applications, pp 283–292.

Inoue T, Furuhashi T, Maeda H and Takaba M (2003). A Proposal of Combined Method of Evolutionary Algorithm and Heuristics for Nurse Scheduling Support System. *IEEE T Ind Electron* 50: 833–838.

Jan A, Yamamoto M and Ohuchi A (2000). Evolutionary Algorithms For Nurse Scheduling Problem. In: *Proceedings of the 2000 Congress on Evolutionary Computation*, pp 196–203.

Kolisch R and Hartmann S (2005). Experimental investigation of heuristics for resource-constrained project scheduling: An update. *Eur J Oper Res*, to appear.

Koop G J (1988). Multiple shift workforce lower bounds. *Manage Sci* 34: 1221–1230.

Li J and Aickelin U (2004). The Application of Bayesian Optimization and Classifier Systems in Nurse Scheduling. *Lect Notes Comput Sc* 3242: 580–589.

Louw M J, Nieuwoudt I and van Vuuren J H (2005). Finding Good Nursing Duty Schedules: a Case Study. *J Scheduling*, to appear.

Maenhout B and Vanhoucke M (2005). An Electromagnetism meta-heuristic for the nurse scheduling problem. *Working paper 05/316*, Ghent University.

Musliu N, Schaerf A and Slany W (2004). Local search for shift design. *Eur J Oper Res* 153: 51–64.

Osogami T and Imai H (2000). Classification of Various Neighbourhood Operations for the Nurse Scheduling Problem. *Lect Notes Comput Sc* 1969: 72–83.

Vanhoucke M and Maenhout B (2005). Characterisation and Generation of Nurse Scheduling Problem Instances. *Working paper series 2005*, Ghent University.

Warner H W (1976). Scheduling Nursing Personnel According to Nursing Preference: A Mathematical Approach. *Oper Res* 24: 842–856.

PERFORMANCE MANAGEMENT

PERFORMANCE ASSESSMENT OF HOSPITALS BY USING SIMULATION

MURAT M. GUNAL and MIKE PIDD
Department of Management Science, The Management School, Lancaster University, UK.

Abstract

It is common to use discrete event simulation (DES) to model healthcare systems for different objectives such as performance improvement, resource planning and facility design. Within this, accident & emergency rooms and outpatient clinics are the most commonly studied. However, despite this apparent success with simulation at an operational level, there remains the challenge of using discrete event simulation in the development and improvement of health policy. We describe the early stages of the development of such a policy-oriented model, aimed at improving performance assessment in the UK National Health Service. As part of the model building, we also discuss simulation software for health care modelling.

Keywords: simulation; health services; hospitals.

Introduction

Health care expenditure has increased massively in the last decade and this trend is likely to continue. Total expenditure on health care in the US in 2004 is estimated at $1.6 trillion (Pierskalla, 2004), having risen at an annual compound rate of 10% in the past decade. In the UK, expenditure on the publicly-funded National Health Service (NHS) is budgeted to rise to over £90 billion by 2007/8. Though technological innovations such as genomics may transform some elements of service delivery, we need to understand their implications for other parts of the patients' pathways though hospitals, so that the benefits can be realised. That is the logistics of health care, which if badly managed, may make it impossible to achieve the benefits promised by scientific advances no matter how much money is spent. This is true in the private health sector and also in the public health sector. In the latter case, tax payers and politicians alike wish to see high quality public services, whilst spending as little as possible.

In the UK, most health care is provided through the publicly-funded NHS in which all services, except some drugs, are provided free at the point of need. It is probably impossible to meet all conceivable demands for health care and so all systems of provision, public or private, include some form of rationing by price or availability. In the UK, the rationing is most evident in waiting lists for treatment, and their reduction has been a major aim of health policy since the Labour Government was elected in 1997. To this end, NHS managers have worked within a performance assessment framework that attempts to measure aspects of the quality of health provision in England. The performance of Acute Hospitals, as one of the major players in the system, has been star-rated in recent years. Excellent performance has been awarded 3-stars and poor performance no stars, the latter hospitals being labelled as failing, with the likely replacement of the Chief Executive. This performance assessment framework has brought improvements in performance together with some suspicions that more could be done if the performance measurement systems were better defined. One way to do this is to develop a simulation model, that represents the entire operations of an acute hospital in sufficient detail for experimentation with different performance assessment regimes.

Since the intended model would represent whole-hospital systems, it would need to cover different levels of detail to be of use in assessing performance measurement at a hospital level. The resulting model is intended to be sufficiently complex to meet these requirements and to offer ease of use.

Simulation In Health Care

Discrete event simulation has been widely used in attempts to improve the delivery of health care (see Fetter and Thompson (1965) for an early example).

Literature surveys

There have been periodic reviews of simulation applications in health care. Examples include Davies (1985), Lehaney and Hlupic (1995), Jun et al (1999) and Fone et al (2003). From these a consistent pattern of successful application emerges, albeit with a focus on specific operational issues such as the management of accident and emergency rooms and of outpatient clinics. Jun et al (op cit) surveys an approximately 30 year period, from the early 1960s, of applications of simulation in healthcare. It reviews 117 journal articles and classifies them according to their stated objectives, focusing on the impact of patient and resource scheduling on patient and work flows; together with the allocation of resources such as beds, rooms and staff. It also concludes that there is a lack of reported studies of more complex, integrated and multi-facility systems. It suggests that the major reasons for this shortage are the level of complexity and the data needs and secondly, the resource requirements, including the time and money needed, to conduct such research. It suggests that the main dilemma in such work is deciding on the appropriate level of detail. Increased detail leads to more realistic representation, which should increase the confidence of stakeholders. However, increased detail requires extensive, validated data and this may be expensive and time-consuming to collect – if indeed it can be collected at all. It further suggests that Soft System Methodologies (SSM) might be used to determine the right level of detail for such complex systems and a similar point is made by Lehaney and Paul (op cit).

Fone et al (2003) is a systematic review of the literature related to the use of simulation modelling in health care and covers almost the same period as Jun et al (op cit). It aims to assess the quality of published studies and to consider their influence on policy rather than on operations. Published work is divided into 5 categories: hospital scheduling and organization, infection and communicable disease, costs of illness and economic evaluation, screening and, finally, miscellaneous. As with Jun et al (op cit), Fone et al (op cit) reports that most of the reported simulation studies focus on very specific aspects of hospitals such as Accident & Emergency rooms, operating theatres, outpatient and inpatient wards. Work aiming to improve scheduling and screening seem to be the most popular areas of research but few papers provide enough detail of model implementation.

Accident & emergency departments and outpatient clinics

It is clear then, that there are many studies of specific hospital departments, of which the most common seem to be Accident and Emergency (A&E), followed by outpatient clinics, operating theatres, laboratories and screening facilities such as MRI. It is interesting to speculate why there are so many simulation models of A&E., such as Miller et al (2004), Sinreich and Marmor (2004) and Shiozaki (2004). One obvious reason is that they are the interface with the hospital for many members of the public and problems with waiting time quickly become public knowledge. A second likely reason though, is that these departments are relatively self-contained and are required to cope with highly variable and exogenous demands for their services. This does not mean that successful A&E simulations are easy to develop, which is an issue discussed in

Carter and Blake (2004), highlighting problems in tracking doctors and in data collection. The same authors discuss some of the issues to be faced when attempting more generic simulation models that might be applied, by suitable parameterization, to different A&E departments.

Studies of outpatient clinics are also popular, which may not be surprising since they have some characteristics that are similar to A&E, even though objectives may differ. In these applications, the focus tends to be on scheduling and capacity planning, as Levy (1989). Lehaney et al (1998) describe in a simulation study of an outpatient department and argue that the use of a graphical interface and visual elements are critical for gaining client's confidence and attention. Kuljis et al (2001) presents a generic outpatient clinics model, CLINSIM, built for UK Department of Health. It simulates how operating policy can influence patient waiting times. Like Lehaney et al (op cit) this also emphasizes the importance of information visualization and iconic animation. The model has been used in 20 clinics in the UK, apparently with some success. Hashimoto and Bell (1996) is another example of a simulation of an outpatient clinic, in this case focusing on staffing and patient scheduling.

Also at clinic level, Swisher et al (2001) model a network of family practice clinics in the US. They first built one clinic model as a template and used it, suitably parameterized, for other clinics. This work is one of the examples of modelling of independent healthcare facilities relying on a common scheduling and information centre.

Hospital level

Though there are many reports of successful simulation studies of A&E departments in individual hospitals, this is not true of simulation studies of whole hospitals. This is probably because hospitals are very complex systems in which there are many components that interact to produce the performance of the hospital. This may be why other approaches, most notably system dynamics, have been used in studying hospitals at a holistic level. Brailsford and Hilton (2001) compare the use of discrete event simulation and system dynamics in health care. However, it must be noted that system dynamics models are not well-suited to detailed modelling and cope rather badly with stochastic variation, which is an important issue in the demand for health care. Brailsford et al (2004) reports on a study of the use of system dynamics to model emergency and on-demand healthcare in Nottingham, UK, which includes a representation of patient flows through different departments in a hospital.

One of the very few hospital level discrete event simulation studies is reported in Moreno et al (2000). This is a simulation of a Spanish hospital intended to predict future hospital behaviour such as waiting times and queue lengths. The idea was to help hospital managers to consider the deployment of resources and the model is, to some degree, linked to the hospital's information system. There are three sub-systems in the overall model; human resources, hospital management and the dynamic model of the hospital. Patient flow resides at the core of the model, which includes a diagrammatic representation of five major types of patient flow: medical, surgical consulting, medical hospitalization, surgical hospitalization and emergency. The model covers not only outpatients and emergency departments but also central services (labs, radiology, haematology, cardiology), wards (medical and surgical), operating theatres, intensive care units and post anaesthetic care units.

DGHPSim

No simulation model is valid for all purposes and so the intended use of the model is a crucial ingredient of any decision about what to include and exclude from the model. The main objective of our research is to develop better understanding of performance measurement in acute hospitals in order to help improve their performance. A known problem with performance measurement is that it can lead to unintended consequences, which include sub-optimization that is, local managers may focus on their own limited and short term goals (Smith, 1995). Therefore for understanding the performance of large complex systems, such as acute hospitals, it makes sense to propose a whole-system model. However, it seems reasonable to ask whether this needs to be a simulation? Would it not be much better to use approximate models, such as those provided by queuing theory? However, taking an analytical approach based on queuing theory is simply too complex if the model is to be sufficiently close to reality to be used in policy making and management. Thus, the scientific core of this work is the development of a whole-system simulation model of a set of generic acute hospitals that, by appropriate parameterization, can be tailored for use in particular acute hospitals.

The model is a generic discrete event simulation model to represent the stochastic demand and resource deployment in a typical district general hospital (DGH). Though the model will be generic, we intend to make it fit particular DGHs by using data available from those hospitals. The model will be based around patient flows through the processes of the hospital and their interaction with resources. This will involve expert judgments.

The broad-brush operation of the model is shown in Figure 1. Two types of demand will be modelled:

1. Exogenous: which will include elective and emergency admissions organized in a specific number of omnibus diagnostic categories.
2. Endogenous: that generated by other hospital processes such as A & E, outpatient and inpatient treatments.

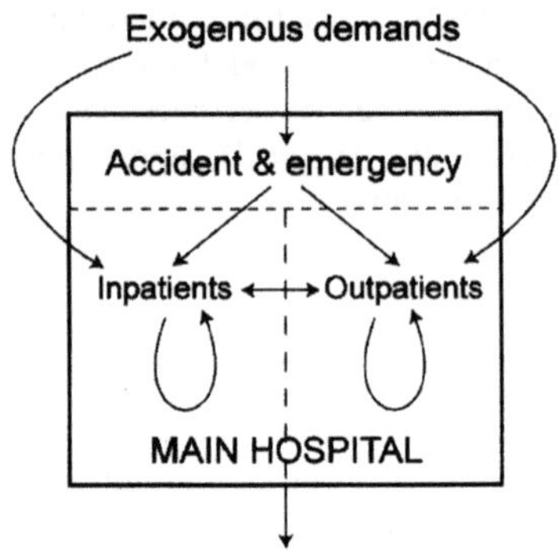

Figure 1: Outline Patient Flows

A typical DGH sees between 50 and 100 A & E patients each day and between 200 and 250 out-patients. It operates between 700 and 800 beds across its specialties and typical lengths of stay for inpatients are between 2 and 8 days. Thus there is considerable opportunity for complexity within the simple flows shown in figure 1. To model endogenous (internally generated) demand will require semi-aggregate data to allow the estimation of state change probabilities for patients in each omnibus diagnostic category as they move through a hospital – that is, on a macroscopic level. Exogenous demand can be modelled as individual patients – that is, on a microscopic level. Hence the model, District General Hospital Performance Simulation (DGHPSim), is a mesoscopic simulator – some elements are modelled in detail (e.g. the continuing flows of patients through A & E), whereas the semi-aggregate state changes of others (e.g. elective admissions each day) can be modelled from probability distributions.

Two model-building approaches seem possible as discussed in Pidd (2004). First, modelling patient flows for each omnibus diagnostic category and following their progress through the hospital, all within a single, unified model. Secondly, modelling each service centre separately (e.g. wards, A&E, clinics, laboratories) and then merging them in a suitable environment. We believe that the latter approach is better to tackle the complex nature of the system than the first approach and will be more appropriate given the varying level of detail across the sub-models.

Hence our work to date has led to the development of an A&E simulator that acts as a generator of emergency inpatient demand as well as producing statistics on the performance of the A&E department itself. This demand, added to the emergency admissions and elective admissions, provides the demand that drives an inpatient model. We have yet to deal with outpatients.

DGHPSim should provide the basis of a scientifically-based, analytical approach with benefits for local managers in district general hospitals, those responsible for NHS performance assessment and the wider research community. DGHPSim is being developed at the level of the individual hospital and hence it is here that its initial benefits will first be realised in the sites selected for our research. The model will be designed to help local hospital managers identify the cause of problems and explore how they may best be resolved. A key task for our research is to assess how the value of models for local use need to be specifically tailored.

DGHPSim is also being developed to improve the bases of performance assessment frameworks by allowing policy staff to:

- Assess the potential systemic impacts of proposed changes to the performance assessment regime, particularly with regard to waiting times;
- Evaluate whether the currently regulatory regime is satisfactory;
- Suggest alternative approaches that would lead to improvement from the perspective of patients;
- Improve regulation of performance by enabling checks on gaming analysis from analyses based on audited data.

Selecting Simulation Software for DGHPSim

Using Commercial Off the Shelf (COTS) simulation software is an option to model DGHs. However the complexity of DGHs and varying levels of detail in the problem definition makes software selection more difficult. The software to be used must be able to cope with both of these modelling process issues, as well as other conventional user and modeller requirements. There are periodic surveys, of which the most comprehensive is that conducted biennially by (Swain 2003) for OR/MS Today, which summarises the features available in simulation packages, with the no specific application area in mind.

Hlupic (2000) surveys academic and industrial simulation software users. It reveals that Simul8™ and Witness™ are the two most commonly used by academics and Witness™ is the most widely used among participant industrial users. After manufacturing, health is the second application area of simulation in academics. The survey also reveals that most of the important features of software are ease of use, ease of communicating with other programming tools and applications for building a user interface. Eldabi and Paul (2001) evaluates three simulation packages for modelling manufacturing systems design with multiple levels of detail and concludes that none of the packages evaluated are well-suited to variable detail modelling. They suggest a methodology for classifying entities with regard to the level of detail of manufacturing system design framework.

In the following section, we examine the use of simulation VIMS for health system logistics. First, selected literature is surveyed to investigate the software packages (VIMS) used to model health systems. Secondly, we briefly summarise our experiences with selected software for modelling an A&E department.

A literature survey

There may be various reasons why a modeller selects a simulation software package to model a problem area. Selection may be based on advice from colleagues, likely costs, personal experience, reading related articles etc. In addition, there are formal approaches to simulation software evaluation and selection, such as that provided in Nikoukaran et al (1998), which presents a hierarchical framework for criteria organized into seven main groups with several subgroups.

Papers in the literature reporting the use of a simulation model are listed in two tables in the Appendix. Table 4 deals with selected papers from a range of journals and Table 5 with selected papers from Winter Simulation Conference (WSC) proceedings. Journal papers cover the period 1999-2005 and includes those with explicit reports of simulation software use. The set of WSC papers consists of health related papers of the period 1997-2004, excluding software tutorials. In total, 20 journal and 20 WSC papers are reviewed and the results are given in Table 1 and Table 2.

The survey reveals that Simul8™ is used in 7 out of 20 selected projects reported in journals. Five of the studies are done with custom-made software rather than using COTS simulation software. In the 20 WSC papers, Arena™ is used in half of the studies and 3 used custom made software. MedModel™ is used in 5 cases overall.

We aimed to include only most recent studies and therefore some papers coincide with the papers reviewed by Jun et al (1999) and Fone et al (2003) literature surveys. A more extensive study can be done by including papers in these surveys.

Table 1: Software used in selected journal papers.

Name	# of used
Simul8	7
Custom software	5
MedModel	2
Arena	1
eM-Plant	1
Excel	1
MicroSaint	1
Modsim	1
XCell+	1

Table 2: Software used in selected WSC papers.

Name	# of used
Arena	10
Custom software	3
MedModel	3
Witness	2
Extend	1
ProModel	1

Our experiences with different software

To check the features of some available packages, we modelled a notional A&E Department of an imaginary hospital using the four packages listed in Table 3.

Table 3: Simulation Software Used.

VIMS Software Name	Version
Micro Saint Sharp ™	*2.0*
MedModel ™	*6.0*
Simul8 ™	*4.0*
Witness ™	*2002*

The A&E process simulated combines and simplifies the features of the A&E departments described in the reference articles. It starts with patient arrivals either by ambulance or directly. Triage and registration is done on the way for ambulance patients, whereas direct arrivals are triaged by a nurse after registration by a clerk. The key node is the doctor's initial examination, since test and treatment needs are decided at this point. Depending on a patient's illness or injury severity, diagnostic tests are ordered, or if this is thought unnecessary the patient is forwarded to nurse treatment. Laboratories and screening facilities are included in the models in a minimum level of detail. Doctor re-evaluation is done after tests and if surgery is required, the patient is forwarded to an operating theatre. At the end of all processes, patients are either "admitted to hospital" or "discharged". Human resources are "Doctors", "Nurses", "Clerks" and they work 8-hour shifts.

Using any of these selected VIMS software it is easy to model such a system. But even though they are similar, there still exist weaknesses and advantageous features. Table 6 shows the comparison of our modelling experience. Main categories are Modelling, Programming, Entity and Resource.

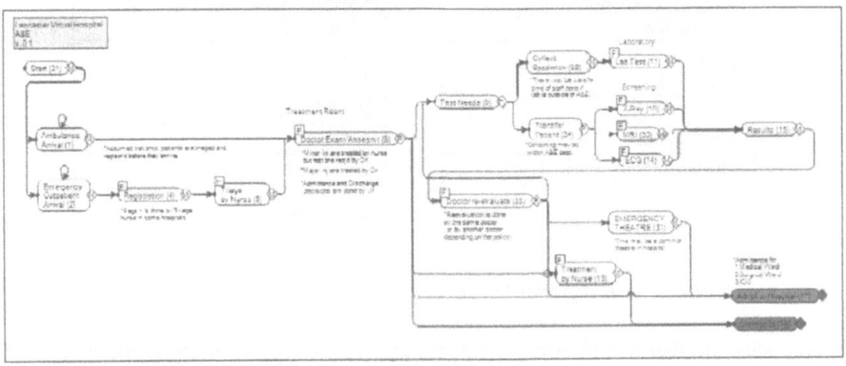

Figure 2: General Flow Chart of assumed A&E Department.
(This chart is drawn using Micro Saint Sharp™ 2.0)

Software selection for the DGHPSim project

It is clear from the literature survey that there seems to be no significant reason, other than convenience, why a particular simulation software has been chosen to model health systems. That is, software choice is not a major issue in these simulations. This could mean that modellers had no problem with their chosen software, or that there were problems but the modellers found some ways to tackle them. The latter sounds more logical and it suggests that perhaps a modeller tailors his or her way of modelling according to the selected software. The best simulation software for a modeller is the one which the modeller knows the best.

Our experience with four different packages was that they offer similar ways to model A&Es. However, A&Es are relatively simple and lead to small models and extending these models to whole hospitals may lead to a different order of problem. However it is difficult to foresee these differences without actually doing the work.

We chose Micro Saint Sharp (2005) for our modelling project for two main reasons; first the modeller is very familiar with it and, second, Micro Saint Sharp uses a general purpose programming language (C#) and access to Microsoft's .NET framework, which gives flexibility in developing complex models and linking them to other software.

Conclusion

It seems sensible to assume that the availability of systems' models, preferable with some generic features, would enable the development of better performance metrics and measurement systems. DGHPSim is being developed with that end in view.

Experience with the selected VIMS packages revealed that each has minor pluses and minuses, but otherwise they are very similar. In all cases, the modelling of simple processes is easy to do whereas, as the complexity increases, limitations appear. An alternative to the use of VIMS software is to write custom software with simulation libraries but the drawbacks for this approach are evident; development time, customization etc. At the time of writing, conceptual modelling is underway and we expect to give more technical account in the future.

Acknowledgments

This research is funded by the Engineering and Physical Sciences Research Council (EPSRC) under grant EP/C010752/1. Research collaborators are Gwyn Bevan & Alec Morton (London School of Economics), Peter C Smith (University of York), Morecambe Bay Hospitals Trust, Royal Orthopaedic Hospitals Trust and the Department of Health.

References

About the NHS. (2005). About the NHS – How the NHS works [online]. Available via <http://www.nhs.uk/England/AboutTheNhs/Default.cmsx> accessed March 14, 2005.

Brailsford S.C. and N.A. Hilton. (2001). A comparison of discrete event simulation and system dynamics for modeling healthcare systems. In: Riley J (Eds.) Proceedings from ORAHS 2000, Glasgow, Scotland, 18–39.

Brailsford S.C., V.A. Lattimer, P. Tarnaras, J.C. Turnbull. (2004). Emergency and on-demand health care: modelling a large complex system. Jnl Opl Res Soc, 55, 34–42.

Carter M.W. and J.T. Blake. (2004). Using simulation in an acute-care hospital: Easier said than done. In: Bradeau M.L., F. Sainfort, W.P. Pierskalla (Eds.) Operations Research and Health Care. Kluwer. Boston.

Davies R. (1985). An assessment of models of a health system. Jnl Opl Res Soc. 1985; 36(8): 679–686.

Department of Health. (2004). National Standards, Local Action. Health and Social Care Standards and Planning Framework 2005/6-2007/8. Department of Health, London.

Eldabi T., R.J. Paul. (2001). Evaluation of tools for modeling manufacturing systems design with multiple levels of detail. International Journal of Flexible Manufacturing Systems. Boston: Apr 2001.Vol.13, Iss. 2; 163.

Fetter R.B. and Thompson J.D. (1965). The simulation of hospital systems. Operations Research. 13, 689–711.

Fone D., S. Hollinghurst, M. Temple, et al. (2003). Systematic Review of the use and value of computer simulation modelling in population health and health care delivery. Journal of Public Health Medicine. 25 (4), 325–35.

Hashimoto F. and S. Bell. (1996). Improving outpatient clinic staffing and scheduling with computer simulation. Jnl General Internal Med. 11, 182–184.

Hlupic V. (2000). Simulation Software: An Operational Research Society Survey of Academic and Industrial Users. Proceedings of the Winter Simulation Conference 2000, Orlando, USA, December 2000, (Eds.) Joines J.A., Barton R.R., Kang K. and Fishwick P., 1676–1683.

Jun J.B., S.H. Jacobson, J.R. Swisher. (1999). Applications of discrete event simulation in health care clinics: a survey. Journal of the Operational Research Society (50) 109–23.

Kuljis J., R.J. Paul, C.M. Chen. (2001). Visualization and Simulation: Two sides of the same coin? Simulation 77 , 3–4, 141–152.

Lehaney B. and V. Hlupic. (1995). Simulation modelling for resource allocation and planning in the health sector. Journal of Royal Society Health. 115, 6, 382–385.

Lehaney B. and R. Paul. (1996). The use of soft systems methodology in the development of a simulation of outpatient services at Watford General Hospital. Journal of the Operational Research Society. (47) 864–870.

Lehaney B., S. Clarke, H. Kogetsidis. (1998). Simulating hospital patient flows. Journal of Royal Society Health. 118, 4, 213–216.

Levy J.L., B.A. Watford, V.T. Owen. (1989). Simulation analysis of an outpatient services facility. Journal of Society Health Systems. 1, 35–49.

Micro Saint Sharp (2005) Micro Analysis and Design [online] http://www.maad.com/index.pl/micro_saint as accessed April 2005.

Miller M.J., D.M. Ferrin, M.G. Messer. (2004). Fixing the Emergency Department: A Transformational Journey With EDSIM. Proceedings of the 2004 Winter Simulation Conference, R .G. In-galls, M. D. Rossetti, J. S. Smith, and B. A. Peters (Eds.), 1988–1993.

Moreno L,, R.M. Aguilar, C.A. Martin, J.D. Pineiro, J.I. Estevez, J.F. Sigut, J.L. Sanchez. (2000). Patient-centered simulation to aid decision-making in hospital management. Simulation. 74, 5, 290–304.

Nikoukaran J., V. Hlupic, R.J.Paul. (1998). Criteria for Simulation Software Evaluation. Proceedings of the 1998 Winter Simulation Conference, (Eds.) D.J. Medeiros, E.F. Watson, J.S. Carson and M.S. Manivannan.

Pierskalla W.P. (2004). Health care delivery. Plenary address to the INFORMS annual meeting, Denver CO, October 2004.

Pidd M. (2004). Computer simulation in management science (5th edition). John Wiley & Sons, Chichester.

Sinreich D. and Y.N. Marmor. (2004). A Simple and Intuitive Simulation Tool for Analyzing Emergency Department Operations. Proceedings of the 2004 Winter Simulation Conference, R .G. Ingalls, M. D. Rossetti, J. S. Smith, and B. A. Peters (Eds.), 1994–2002.

Smith P.C. (1995). On the unintended consequences of publishing performance data in the public sector. Intl. Journal of Public Administration. 18, 2&3, 277–310.

Swain J. (2003). Simulation Software Survey. OR/MS Today. August 2003. http://www.lionhrtpub.com/orms/surveys/Simulation/Simulation.html

Swisher J.R., S.H. Jacobson, J.B. Jun, O. Balci. (2001). Modelling and Analyzing a Physician Clinic Environment Using Discrete-event (visual) Simulation. Computers & Operations Research. 28, 105–125.

Takakuwa S. and H. Shiozaki. (2004). Functional Analysis for Operating Emergency Department of a General Hospital. Proceedings of the 2004 Winter Simulation Conference, (Eds.) R .G. Ingalls, M. D. Rossetti, J. S. Smith, and B. A. Peters, 2003-2011.

Appendix

Table 4: Selected Journal Papers

No	Reference	Software Used	Place of Modelling	Purpose of Study
1	Ledlow G.R., Bradshaw D.M., Perry M.J. (1999), Animated Simulation: a Valuable Decision Support Tool for Practice Improvement. Journal of Healthcare Management 44-2.	MedModel	Clinic	Design
2	Eldabi T., Paul R.J., Taylor S.J.E. (1999) Computer Simulation in Healthcare Decision Making. Computers & Industrial Engineering 37, 235–238	Simul8	Clinic	Breast cancer treatment trial
3	Lehaney B, Clarke SA, Paul RJ (1999) A case of an intervention in an outpatients department. J Ops Res Soc, 50, 877–891	Simul8	Derma-tology Outpatient Clinic	Waiting times
4	Kim SC, I Horowitz, K.K. Young, TA. Buckley (1999) Analysis of capacity management of the intensive care unit in a hospital. European Journal of Operational Research 115, 36–46.	XCELL+	ICU	Capacity utiliza-tion
5	Bagust A., M. Place and J.W Posnett (1999) Dynamics of bed use in accommodating emergency admissions: stochastic simulation model. BMJ 1999;319;155–158.	Excel	A&E	Waiting times
6	Moreno L,, R.M. Aguilar, C.A. Martin, J.D. Pineiro, J.I. Estevez, J.F. Sigut, J.L. Sanchez. (2000) Patient-centered simulation to aid decision-making in hospital management. Simulation. 74, 5, 290-304.	MODSIM Modula2 based language	Hospital	General
7	Kuljis J., Paul RJ., Chen CM. (2001) Visualization and Simulation: Two sides of the same coin? Simulation 77 (3-4) 141-152	Custom Software CLINSIM	Generic Clinic	General

8	Pulat PS, Kasap S, Splinter GL (2001). Simulation study of an ideal primary care delivery system. Simulation 76 (2): 78-86	Arena	Hospital	Bed capacity planning
9	Swisher J.R., Jacobson S.H., Jun J.B, Balci O. (2001). Modelling and Analyzing a Physician Clinic Environment Using Discrete-event (visual) Simulation. Computers & Operations Research 28, 105}125	Custom Software with Visual Simulation Environ-ment	Clinics	General
10	Groothuis S., van Merode G.G., Hasman A. (2001) Simulation as decision tool for capacity planning. Computer Methods and Programs in Biomedicine 66, 139-151.	MedModel	Cardiac Dept.	Capacity planning
11	Harper PR, AK Shahani (2002) Modelling for the planning and management of bed capacities in hospitals. Jnl Ops Res.Soc. 53, 11-18.	Custom Software TOCHSIM	Hospital	Bed capacity
12	Couchman A., Jones D.I., Griffiths K.D. (2002) Predicting the future performance of a clinical biochemistry laboratory by computer simulation. Simulation Modelling Practice and Theory 10, 473–495	Simul8	Bio-chemistry Laboratory	Capacity planning
13	Harper P.R., H.M. Gamlin (2003) Reduced outpatient waiting times with improved appointment scheduling: a simulation modelling approach. OR Spectrum, 25: 207–222.	Simul8	ENT Clinic	Appoint-ment Sched-uling
14	Syi Sua, Chung-Liang Shihb (2003), Modeling an emergency medical services system using computer simulation. International Journal of Medical Informatics 72, 57–72	eM-Plant	A&E	Emer-gency response policy
15	Baldwin L. P.,T. Eldabi, R.J. Paul (2004) Simulation in healthcare management: a soft approach (MAPIU). Simulation Modelling Practice and Theory, 12 541–557.	Simul8	Clinic	Liver trans-planta-tion
16	Griffiths JD, N Price-Lloyd, M Smithies, JE Williams (2005) Modelling the requirement for supple-mentary nurses in an intensive care unit. Jnl Ops Res Soc., 56, 126–133.	Simul8	ICU	Nurse scheduling and number of nurses required.

17	Ashton R., L Hague, M Brandreth, D Worthington, S Cropper (2005) A simulation-based study of a NHS Walk-in Centre. Jnl Ops Res Soc., 56, 153–161.	Micro Saint	Walk in Centre	Design and plan of operations of a Walk in centre
18	Rauner MS, SC Brailsford, S Flessa (2005) Use of discrete-event simulation to evaluate strategies for the prevention of mother-to-child transmission of HIV in developing countries. Jnl Ops Res Soc., 56, 222–233	Custom Software with patient-oriented simulation technique (POST)	Disease Prevention	HIV prevention
19	Van der Meer RB, LA Rymaszewski, H Findlay, J Curran (2005) Using OR to support the development of an integrated musculo-skeletal service. Jnl Ops Res Soc., 56, 162–172	Simul8	Musculo-skeletal service in a hospital	Waiting times
20	Harper PR, S Phillips, JE Gallagher (2005) Geographical simulation modelling for the regional planning of oral and maxillofacial surgery across London. Jnl Ops Res Soc., 56, 134–143.	Custom Software with TOCHSIM library	London area	Oral and maxillo-facial surgery planning & scheduling

Table 5: Selected Winter Simulation Conference Papers.

No	Reference	Software Used	Place of Modelling	Purpose of Study
1	Swisher J.R., B. Jun, S.H. Jacobson, O.Balci (1997). Simulation of the Queston Physician Network. Proceedings of the 1997 Winter Simulation Conference.	Custom Software Visual Simulation Environment	Clinic	General
2	Pitt M. (1997). A Generalised Simulation System to Support Strategic Resource Planning in Healthcare. Proceedings of the 1997 Winter Simulation Conference.	Witness + VB interface	Hospital	Bed size planning (PRISM project is funded by EPSRC)

3	Barnes C.D., J.L.Quiason, C.Benson, D.McGuiness. (1997). Success stories in simulation in health care. Proceedings of the 1997 Winter Simulation Conference.	Witness	Outpatient clinic	pre-op proce-dures, space utiliza-tion and out-patient studies
4	Johnson W.C. (1998). Birth of a new maternity process. Proceedings of the 1998 Winter Simulation Conference.	MedModel	Maternity ward	Capacity planning
5	Rossetti M.D., Trzcinski G.F., Syverud S.A. (1999). Emergency Department simulation and deter-mination of optimal attending physician staffing schedules. Proceedings of the 1999 Winter Simulation Conference.	Arena	A&E	Staffing, sched-uling.
6	Lowery J.C., J.A. Davies. (1999). Determination of operating room requirements using simulation. Proceedings of the 1999 Winter Simulation Conference.	MedModel	Operational Theatre	Size, sched-uling.
7	Cahill W., Render M. (1999). Dynamic simulation modelling of ICU bed availability. Proceedings of the 1999 Winter Simulation Conference.	Arena	ICU	Bed size planning
8	Weng M.L., A.A. Houshmand. (1999). Healthcare simulation: A case study at a local clinic. Proceedings of the 1999 Winter Simulation Conference.	Arena	Clinic	A new perfor-mance measure, cash flow.
9	Alvarez A.M., Centeno M.A. (1999). Enhancing simulation models for emergency rooms using VBA. Proceedings of the 1999 Winter Simulation Conference.	Arena	A&E	General

10	Ramis F.J., Palma J.L., Estrada V.F., Coscolla G. (2002). A simulator to improve patient's service in a network of clinic laboratories. Proceedings of the 2002 Winter Simulation Conference.	Arena + Excel	Laboratory	Standard-ize the service processes, the assign-ment of person-nel and to guide invest-ment decisions
11	Eleazer M., Gronhaug R., Haugene K. (2003). Proposals to reduce over-crowding, lengthy stays and improve patient care: Study of the geriatric department in Norway's largest hospital. Proceedings of the 2003 Winter Simulation Conference.	ProModel	Geriatric Department	Bed size, LOS.
12	Blasak R.E., Armel W.S., Starks D.W., Hayduk M.C. (2003). The use of simulation to evaluate hospital operations between the emergency department and a medical telemetry unit. Proceedings of the 2003 Winter Simulation Conference.	Arena	A&E	Waiting time
13	Wong C., Geiger G., Derman Y.D., Busby C.R., Carter M.W. (2003). Redesigning the medication ordering, dispensing and administration process in an acute care academic health sciences centre. Proceedings of the 2003 Winter Simulation Conference.	MedModel	Hospital	Medica-tion ordering, dispen-sing
14	Baesler F.F., H.E. Jahnsen, M. DaCosta (2003) The use of simulation and design of experiments for estimating maximum capacity in an emergency room. Proceedings of the 2003 Winter Simulation Conference.	Arena	A&E	Bed and staff capacity
15	Miller M.J., Ferrin D.M., Szymanski J.M. (2003) Simulating six sigma improvement ideas for hospital emergency department. Proceedings of the 2003 Winter Simulation Conference.	Extend + Visio + Access	A&E	Process improve ments

16	Centeno M.A., Giachetti R., Linn R., Ismail A.M. (2003) A simulation – ILP based tool for scheduling ER staff. Proceedings of the 2003 Winter Simulation Conference.	Arena	A&E	Number of staff required, sched-uling
17	Guo M., Wagner M., West C. (2004) Outpatient clinic scheduling – a simulation approach. Proceedings of the 2004 Winter Simulation Conference.	Arena	Outpatient Clinic	Appoint-ment sched-uling
18	Miller M.J., Ferrin D.M., Messer M.G. (2004) Fixing the emergency department: A transformational journey with EDSim. Proceedings of the 2004 Winter Simulation Conference.	Custom Software EDSim	A&E	General
19	Takakuwa S., Shiozaki H. (2004) Functional analysis for operating emergency department of a general hospital. Proceedings of the 2004 Winter Simulation Conference.	Arena	A&E	Waiting times, patient flows
20	Sinreich D., Marmor Y.N. (2004) A simple and intuitive simulation tool for analyzing emergency department operations. Proceedings of the 2004 Winter Simulation Conference.	Custom Software	A&E	General *A generic A&E model-ling study.

Table 6: Comparison of selected simulation software.

	Micro Saint Sharp™ v2	MedModel™ v4	Simul8™ v4	Witness™ 2002
Modeling				
Object types	Task, sub-networks, groups (Queues are attached to Tasks.)	Locations, Entities, Path Networks, Resources,	Work centers, work entry&exit points, storage (queues), conveyers, resources.	Machines, buffers, labor, conveyor etc... (There are more items in Designer Elements palette)
Animation-2D	Iconize, after the network is built	Iconize, as the model is built	Iconize, as the model is built	Iconize, as the model is built
Diagrammatic Process Flow	Yes, Easy to create	Yes, With icons.	Yes, With icons.	Yes, Process view can be seen
Warm-up Period	No, It must be coded explicitly	Yes, It can be specified	Yes, It can be specified	Yes, It can be specified
Simulation Speed	Very fast, adjustable	Very fast, adjustable	Very fast, adjustable	Very fast, adjustable
Data Collection	Via snapshots, charts, printOut function	Automatically (As time series or basic)	Automatically	Automatic-ally
Run length	It must be coded explicitly (e.g.if clock==1440 then Model.Halt();)	Yes, It can be specified	Yes, Can be set as "Results collection period"	Yes, "Stop time at..." box.
Queue Management	Queues are attached to Tasks, different disciplines can be selected.	Queues are attached to "Locations". Different disciplines can be given (FIFO, LIFO, order by att.)	Queue capacity, shelf life can be specified.	Buffers are manageable objects
Arrivals	No, Can be defined only by coding. (from "Timing tab" of generator-loop task)	Yes, Can be defined by "Arrival Cycle Editor"	Yes, Different arrival distributions for different time slots can be given.	Yes, Different arrival rates can be entered by input text files.

Programming				
Language	C#	Logic Elements	Visual Logic	Actions and rules (BASIC)
Need	Yes, Must write code to gain full advantage (e.g. Shift patterns, arrival patterns)	Yes, Resource use, parameters to statistical distributions, variable, graphic changes must be coded. (e.g. USE DR ,20 for E(20))	Yes, Basic things can be done without coding. (e.g.No need to code for shift&arrival patterns)	Yes, Build in functions can be used to do things.
Communication with Windows Applications	It can read/write from/to Excel, web service, ADO source and text.	Arrays can be imported from Excel. Data plots can be imported to Excel.	It can read Excel files, dynamically linked.	It can link Excel spreadsheets, read and write data to/from.
Global variables	Yes, Data types: Integer, floating, Boolean, Entity, arbitrary, string, object, hashtable.	Yes, Data types: Integer, real	Yes, Data types: Num., Text, Time, Spreadsheet (can copy/pasteto/from Excel)	Yes, As Designer Elements Data types: Integer, real, name, string.
Arrays	Yes, All data types can be defined as arrays.	Yes, Integer, real types. Can be exported or imported to/from Excel	Yes, As defined in Spreadsheet variable type.	Yes, All data types can be defined as arrays.
Entity	Entities are "Patients" and "Tests"			
Name	Entity	Entity	Work Item	Parts
Attributes	Yes, Data types: String, number, boolean, object, array, entity	Yes, Data types: Integer, real	Yes, Data types: Text and number	Yes, Data types: Integer, real, name, string

Types	One entity type	One entity type exists, but different names can be given to create different types of entities.	As many entity type as you want	As many entity type as you want
After processing an entity	The same entity leaves the task	A different type of entity can leave	A different type of entity can leave (e.g. a test request may leave after doctor treatment of a patient)	A different type of entity can leave
Resource	Resources are "Doctors", "Nurses", "Clerks".			
Ease of usage	Coding needed (e.g. "number of Doctors--;")	No coding needed, (done via GUI, points must be specified where resources will be used)	No coding needed, (done via GUI)	No coding needed, (just typing the name of resource)
Shift assignment	No shortcut for shifts. Coding is needed.	Yes, Easy to define shifts by "Shift Editor"	Yes, Easy to define shifts by "Shift Patterns" editor.	Yes, Easy to define.
"Process efficiency" and "Downtime" entry	No, It must be coded explicitly	Yes, Downtimes can be set.	Yes, Detailed efficiency for the processes can be specified.(e.g. breakdowns & recoveries)	Yes, Downtimes, setup times can be set.

AN EVALUATION OF THE MANAGED CARE PROGRAM AS APPLIED TO THE BRAZILIAN SUPPLEMENTARY HEALTH SERVICE

LÉA MARIA DANTAS SAMPAIO, MARIO JORGE FERREIRA DE OLIVEIRA and ANÍBAL ALBERTO VILCAPOMA IGNÁCIO
Federal University of Rio de Janeiro, Brazil.

Abstract

This paper focuses on the economic and financial performance of supplementary health operators in the Brazilian market. A large database from the National Supplementary Health agency (ANS) provides the necessary background for the study. A Data Envelopment Analysis (DEA) model is proposed to evaluate the performance and to rank a sample of operators. Three companies that apply the Managed Care Program (MCP) are chosen and evaluated in accordance with the relative efficiency score. It is expected to identify the necessary reform strategies of the sector in order to suggest a new-fangled supplementary health assistance program.

Keywords: data envelopment analysis; managed care program; supplementary health programs.

Introduction

The accessibility to health services in Brazil is a big issue, and much remains to be done to improve the quality of services. There is, from one side a huge and unified public health system (SUS), which is not able to cope with the increasing demand. From the other side there are a number of supplementary health operators, who provides better services to about 22% of the population that can afford to pay for private health. The main concern is the imbalance of health service level provided by the supplementary operators, and a national agency (ANS, 2005) has been created to regulate the sector.

The Managed Care Concept

A review of the literature reveals that Managed Care (MC) is a concept that started to be developed in the early seventies in United States of America. The Health Maintenance Organization (HMO) act focuses on management policies with the aim reducing the costs of health delivery (Andrade & Lisboa, 2000). Some of the outcomes are:

- The establishment of a healthy competition among health plans;
- The reduction of the growing expenses with health provision;
- The rationalization of medical and hospital services;
- The sharing of risk between the entities involved.

The MC programs are composed of at least four entities:

- Insurance companies or managed organizations that finance and manage the services;
- Organized groups of providers that are basically medical and hospital associations;
- Independent providers that are basically doctors;
- Patients. The introduction of new plans makes it possible to adjust the supply and the demand.

The MC, however defined, is a health care system that is designed to control costs through managed programs in which the physician accepts constraints on the amount charged for medical care and the patient is limited in the choice of a physician. One of the issues that arise for MC programs is that the creation of several health plans makes complex contractual relation between the agents.

MC can also be considered as an organized effort by health insurance plans and providers to use financial incentives and organizational arrangements to alter provider and patient behaviour so that healthcare services are delivered and utilized in a more efficient and lower cost manner (Williams & Torrens, 1999).

The expansion of the MC principles is the result of its positive effect on the capacity to improve the quality of the health delivery. There are reports of cost reduction and coverage improvement in the Medicaid and Medicare public programs (Médici, 2003). Since the 1990's, American Managed Care Organizations (MCO) have been looking for new markets outside USA. The MCO executives have forecast large investment returns in Latin America. Local governments usually justify the adoption of multinational investments in MC under the flag of management inefficiency of the public sector (Iriart *et al.*, 2001).

The MCO´s strategy to enter the Brazilian market is usually to approach already established pre-paid health plan companies. The establishment of joint ventures with local health insurances companies is also common. MC techniques encompass a range of financial incentives and arrangements. However, these techniques are often imported without empirical evidence of their effectiveness for developing countries with constrained economic resources (Luck & Peabody, 2002a; 2002b).

The Evolution of the Brazilian Health Sector

The first initiatives to structure the health sector in Brazil started between 1920 and 1940. At that time, the social rights were dependent on the position of individuals in the official labour market. There was a social security system defined by the government. This system involved retirement and pension institutes, which provided different levels of medical assistance to people in the labour market.

The supplementary health service emerged between 1940 and 1960. The first initiatives happened in both the public and private sectors. The first private health plans were commercialized in the mid-1960s. Two classes of operators arise: the group medicine, that is essentially a network of hospital owners, and the medical co-operatives that are private enterprises, created by doctors to compete with group medicine. Both classes did deliver their own health service, attempting to capture the market share vacated by the National Institute of Medical Care for Social Welfare (INAMPS) (Bahia & Viana, 2002).

A health sector reform was proposed from 1980. The social and political forces committed themselves to changes in the health sector and interacted through meetings and debates on a national scale. Preceded by a number of minor public health meetings, the 8th National Health Conference was held in 1986. With an explicit political character, the health reform issues and concepts were then consolidated and put into practice at the National Constitution of 1988.

The technical/social welfare reform plan institutionalized health as a social right, proposing the establishment of an egalitarian SUS providing universal access to any health care service. The services were to be paid for by public funds, providing services at all levels. The new SUS was conceived to promote the hegemony of the public sector. It proposed the decentralization and strengthening of local authorities, the regulation and control of the private sector and social participation (Collins *et al.*, 2000).

As a result of the government's inability to implement the SUS ideals, a huge market for private health plans was recognized by the end of the 1980s. It was followed by the intensification of individual health plan sales, the decisive introduction of large-scale insurers and the subscription to supplementary medical assistance, by new groups of workers. Introduced in 1994, the plan to stabilize the Brazilian currency provided greater purchasing power, generating greater sales of individual health plans including charity hospitals and medical companies from different regions.

The Federal Senate passed, in 1997, a draft law regulating the supplementary health segment. The project underwent, in 1998, a major alteration that involved broadening the reach of the regulations to health assistance, by means of a more specific judgment of the available plans, including rigorous rules protecting the customer's interests. A second amendment to the regulation was made in 1999. It resulted in the unification of all areas regulated by the Ministry of Health. From this unification followed the ANS institution and the transfer of the regulatory powers to this agency in 2000.

The ANS is then a regulating agency subordinate to the Ministry of Health, which promotes the equilibrium in the relationship between several segments aimed at building, together with the society, a solid, socially oriented market. The goal is to promote the public interest in the supplementary health assistance, regulate the operators- inclusive as far as their relationship with providers and consumers are concerned- and contribute to the development of the health services in the country (ANS, 2005).

The Brazilian Supplementary Health Segment (SHS)

The Brazilian SHS involves more than 2000 companies operating health plans, several thousand doctors and other professionals, hospitals, labs and clinics. This network provides a service to about 40 million consumers that use private health plans, which include consultations, medical examinations and hospitalization (ANS, 2005).

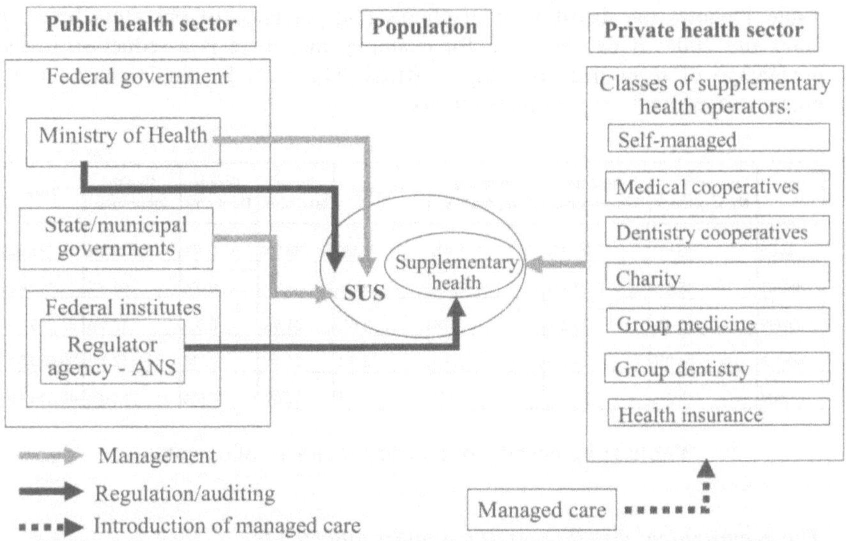

Figure 1: The structure of the public and private health sectors in Brazil.

Figure 1 shows the structure of both the Brazilian public and private health sectors. The federal government is responsible for the management and regulation of public health. The government regulates, through the ANS, the SHS.

The operators are distributed in seven classes: group medicine, group dentistry, charity, medical and dentistry cooperatives, self-management and health insurance companies. It is noticeable that the ANS regulates the segment, but the management of the SHS is left to the individual operators. Some American managed care organizations are interested in the Brazilian private health sector. Most of them are connected with the health insurance. Three companies[1] that apply the Managed Care Program (MCP) are evaluated, in this paper, according to their relative financial performance:

- HI1 - health insurance and MCO1;
- HI2 - health insurance and MCO2;
- HI3 - health insurance and MCO3.

[1] Because of confidentiality reasons, the actual names of the operators are not mentioned.

Table 1 shows the distribution of clients (%) per class of operators between 2000 and 2004. It can be seen, for example, that there is a reduction in the percentage of insurance company's clients. There is also an increase in the number of medical cooperative's clients.

	Self Management	Medical Cooperative.	Dentistry Cooperative.	Charity	Group Medicine	Group Dentistry	Health Insurance	Total
jun/00	15.8%	22.4%	1.9%	3.5%	35.5%	5.1%	15.8%	37,731,819
dez/01	15.6%	23.7%	2.4%	3.6%	33.8%	6.0%	14.9%	37,412,466
dez/02	15.4%	24.3%	2.9%	3.6%	34.4%	6.9%	12.5%	37,859,334
dez/03	14.9%	24.7%	3.2%	3.3%	34.1%	7.7%	12.1%	39,025,546
dez/04	13.9%	25.1%	3.6%	3.2%	33.6%	8.9%	11.8%	40,316,717

Table 1: Percentage of clients per classes of operators.

The geographical distribution of operators and clients

MC programs require a range of financial incentives and organizational arrangements. An adequate demand and population density are important preconditions (Luck & Peabody, 2002a). A previous study (Fernandes *et al.*, 2003) discusses important issues related to the SHS market based on a project involving the ANS and United Nations (ANS, 2005). An evaluation of the geographical distribution of the operators and clients is made. A large database from the National Supplementary Health agency (ANS) provides the necessary background for the study. Figure 2 shows the distribution of clients per state and the class of operators.

It can be seen that the distribution of operators and clients is concentrated in the southeast region of the country. It is noticeable that this is the most populated and industrialized region of the country. Furthermore, the wage concentration in the states of Minas Gerais (MG), Espírito Santo (ES), Rio de Janeiro (RJ) and São Paulo (SP) is certainly higher than in the other regions.

Another important issue is that the Brazilian SHS is extremely unbalanced. Figure 3 shows the concentration of clients per number of operators. It can be seen that 90% of the clients are connected to about 28% of the total number of operators.

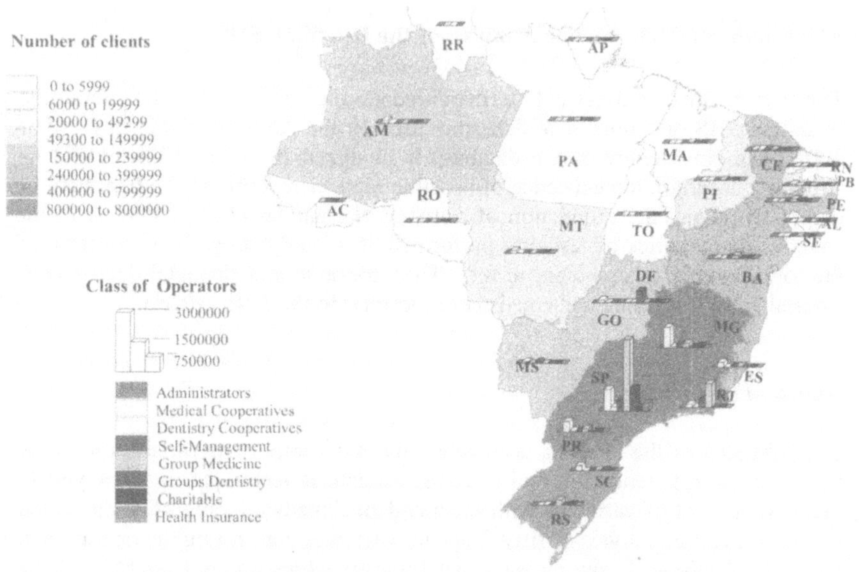

Figure 2: Distribution of clients and operators per state

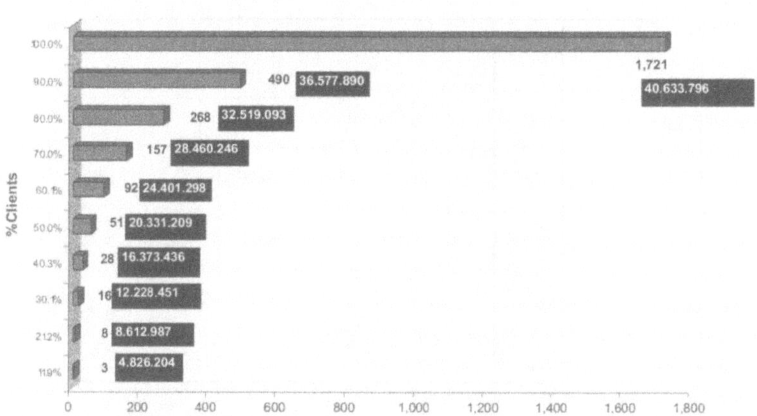

Figure 3 - Concentration of clients per number of Operators:ANS-2004.

The Financial Performance Analysis of the Brazilian SHS

The performance analysis is based upon economic and financial data about the Brazilian SHS operators. The data, provided by the ANS, is related to the year 2002. The consistency of the database is analyzed in order to understand its structure, ascertain the existence of possible register repetitions, the treatment of empty fields and the production of statistics (Fernandes *et al.*, 2003). A factor analysis and a series of tests are performed. The study suggests a reduction of the total number of operators to 897. The economic and financial data is considered valid if it is deemed consistent according to the ANS criteria.

Selection of Variables

The first step of the analysis is to select the most important variables to study the operator efficiency. A multivariable statistical technique is employed to identify of a set of variables with independent dimensions. Figure 4 shows the result of a factor analysis of fifty financial variables concerning the operators in the ANS database. It can be seen that three variables account for 88% of the database variance. Cash/banks, share capital and the premiums received are the most important factors that compose the balance sheet.

Cash/banks

Shared Capital

Premiums received

Rotated Component Matrix				
	Component			
	1	2	3	4
VL ATVO CIRC DISL	**0.837**			
VL ATVO IMOV	0.521	0.691		
VL ATVO PERM BENS	0.829			
VL ATVO CIRC PECA			0.586	-0.529
VL ATVO PERM DEPR		0.589		
VL [illegible] PREJ		-0.825		
VL PASS PATR CAPT		**0.917**		
VL PASS DEBT DIVR			0.841	
VL DEMO DESP ADMT	-0.791			
VL DEMO COPR LIQU	0.663			-0.509
VL DEMO FINC DESP			-0.821	
VL DEMO OUTR RECT				**0.893**
VL DEMO EVEN	-0.654			0.507
VL DEMO FINC RECT	0.528			

Extraction Method: Principal Component Analysis.
Rotation Method: Varimax with Kaiser Normalization.

Figure 4: Factor analysis result.

Data Envelopment Analysis (DEA)

The DEA approach was developed by Charnes *et al.* (1978) to determine the efficiency of productive units. In the present case, DEA allows one to evaluate the relative efficiency of each operator as the assessed Decision Making Unit (DMU) and rank them, taking into account the resources at their disposal (inputs) and the results achieved (output). The objective of the analysis is to compare a certain number of DMU that accomplish similar responsibilities and diverge in the amounts of inputs they consume and outputs they generate. There are two classic DEA models: the CRS or CCR model (Charnes *et al.*, 1978), which considers constant returns to scale and the VRS or BCC model (Banker *et al.*, 1984) that considers variable returns to scale. The former model does not assume proportionality between inputs and outputs.

The DEA models allow one to measure and locate the inefficiency, based on a linear piecewise frontier production function that identifies the benchmark for inefficient DMU. This reference point is determined by the projection of inefficient unities over the efficiency frontier. There are two basic orientations for projections according to basic models: input (when one wants to minimize the resources, keeping the outputs constant) and output orientation (when one wants to maximize the outputs without reducing the inputs). It has decided to use, in this paper, an output oriented BCC model. The inputs of the model are cash/banks and share capital, and the outputs are number of clients and premiums).

The DEA model

The usual VRS linear programming (LP) formulation for the DEA model suggested in the existing literature is:

$$\text{Min } h'_0 = \sum_{i=1}^{m} v(i) * x(i,o) + v0$$

Subject to:

$$\sum_{r=1}^{s} u(r) * y(r,o) = 1$$, (The "o" represents the observed DMU)

$$\sum_{r=1}^{S} u(r) * y(r,j) - \sum_{i=1}^{m} v(i) * x(i,j) + v0 \leq 0 \quad j = 1 \ldots n;$$

$$u(r) \geq 0 \quad r = 1 \ldots s;$$

$$v(i) \geq 0 \; ; i = 1 \ldots m.$$

The x(i,j) represent the inputs "i" of the DMU "j"; y(r,j) represent the output "i" of the DMU "j"; v(i) and u(r) represent, respectively, the weights given to the inputs "i" and the outputs "r"; "v_0" is a scale factor (if positive, it indicates that DMU operates with increasing returns to scale; if negative, it indicates decreased returns to scale) (Cooper *et al.*, 2000). According with the basic model, for each DMU under analysis, the efficiency is given by h_o^{-1}, the inverse of the objective function. If h_o^{-1} is equal to 1, the DMU under analysis is considered efficient. The decision variables of the LP formulation are v(i), u(r) and v_0.

In some cases, the basic model shows some distortion. It means that if weights v(i) or u(r) given to a particular "efficient" DMU are equal to zero, then the DMU are "pseudo-efficient". In this article an extension of the VRS basic model has been used to avoid this possible misrepresentation. The chosen model, named BCC non Archimedean restricts the weights v(i) and u(r) to an infinitesimal element ε greater that zero. Therefore the modification in the basic model is:

$$u(r) \geq \varepsilon \; r = 1 \ldots s;$$
$$v(i) \geq \varepsilon \, ; i = 1 \ldots m.$$

The solution of any DEA models that involves the non-Archimedean elements requires the choice of an adequate range value for ε. According to the literature (Jahanshahloo & Khodabakhshi, 2004), the appropriate value of ε should be determined to satisfy the assurance interval (0, ε*) given by the expression $\varepsilon^* \leq (1/\max \sum y(r,j))$.

The Results

The result of the DEA model is the ranking of the financial efficiency score of the 897 SHS operators. The scores vary between zero and one. The score one means 100% efficiency. Three groups of companies are created, depending on the efficiency score level, with the help local experts. The high level, which is the efficient frontier group comprises companies with 100% efficiency. The medium level includes the companies with efficiency greater than or equal to 50%, but less than 100%. The others are classified in the low level.Table 2 shows the percentage of operators for each level. It also shows, for each level, the number and percentage of input and output. One can see, for example that only 4% of the operators are considered "efficient".

Score level	All of the operators classes	Cash/banks	Share capital	Clients	Premiums received	Number of Operators
High	Total	106.272.733	95.521.067	4.298.231	2.713.919.639	
	% of Total variables	4%	3%	18%	24%	4%
Medium	Total	1.462.933.104	1.224.773.857	8.371.595	4.281.693.079	
	% of Total variables	54%	45%	35%	37%	7%
Low	Total	1.165.003.068	1.412.449.173	11.087.719	4.459.452.224	
	% of Total variables	43%	52%	47%	39%	89%

Table 2: Result of the analysis.

The main focus of this paper is to evaluate the performance of operators that are involved with MC programs. Therefore, we shall concentrate on the efficiency score of companies that apply such programs. As mentioned before, at least three companies (HI1, HI2 and HI3) have this profile. Table 3 shows the input and output data from three Brazilian operators that have a joint venture with MCO.

Operators	Cash/banks	Share capital	Clients	Premiums received
HI1	671.603.198	398.386.025	1.588.691	647.204.825
HI2	34.820.367	20.000.000	57.142	22.345.447
HI3	539.052.026	346.000.000	2.322.514	572.195.691
Total	1.245.475.592	764.386.025	3.968.347	1.241.745.963
Other operators	2.734.208.905	2.732.744.097	23.757.545	11.455.064.942
Total /Other operators	46%	28%	17%	11%

Table 3: Comparative data for the MCO operators.

It can be argued that the three operators appear not to be doing well, as far as the relative efficiency score is concerned. As an example, one can see that the three companies together concentrate 46% of the total cash/banks of the sample. However, the number of clients and the premiums received do not follow the same proportion. As shown in Table 2, the most efficient operators together hold only 4% of the cash banks input and produce better outcomes.

In order to complement the analysis, it is interesting to see the positions of the three operators in the efficiency score rank. Figure 6 illustrates that not all of the operators connected with MCO are well positioned in the rank. Two of them fall in the middle efficiency range with 70% and 59% score, respectively. The third company shows a very low (6%) score.

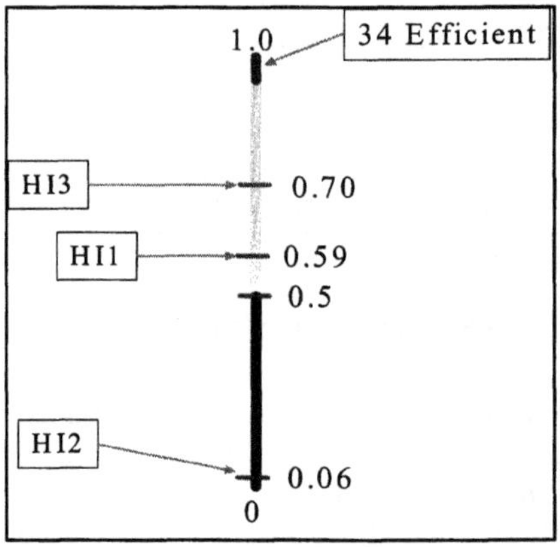

Figure 6: Placement of the Scored Managed Care in Brazil.

Conclusions

The economic and financial performance of a sample of supplementary health operators has been evaluated in this paper. The study entails historical aspects of health management in Brazil and the analysis of the structure of the public and private health assistance. It is shown that the distribution of operators, and clients throughout the country is very unequal, pointing to asymmetric geographical aspects of the Brazilian SHS.

The analysis reveals that the great majority of the studied organizations do not seem to be operating efficiently. The assistance quality, however defined, does not correspond to the high price and to the set of constraints imposed by the organizations. 47% of clients belong to the most inefficient operators. Great social problems could occur if they decline or go bankrupt.

The entry of the managed care programs into the Brazilian market has stimulated the inequality already existing in the supplementary health segment. The efficiency score of the three insurance companies that have made a joint venture with MCO is assessed. Their relative efficiency score is not as expected in terms of the management of resources for the production of results.

References

Agência Nacional de Saúde Suplementar (ANS), *Contrato de gestão.* Available at: http://www.ans.gov.br

Andrade, M. V. & Lisboa, M. B. 2000, 'Sistema privado de seguro-saúde: lições do caso americano'. *Revista Brasileira de Economia,* vol. 1, no. 54, pp. 5–36.

Bahia, L.E. & Viana, A. L. D. 2002, 'Regulação & saúde: estrutura, evolução e perspectivas da assistência médica suplementar', introdution in: *ANS/ MS*, ed. Rio de Janeiro, Brazil.

Banker, R. D., Charnes, A. & Cooper, W. W. 1984, 'Some models for estimating technical scale inefficiencies in data envelopment analysis', *Management Science,* vol. 30 no. 9, pp. 1078–1092.

Charnes, A., Cooper, W. W. & Rhodes, E. 1978, 'Measuring the efficiency of decision-making units'. *European Journal of Operational Research*, vol. 2, pp. 429–444.

Collins, C., Araújo, J. & Barbosa, J. 2000, 'Decentralising the health sector: issues in Brazil', *Health Policy,* vol. 52, pp. 113–127.

Cooper, W. W., Seiford, L. M. & Tone, K. 2000, 'Data envelopment analysis: a comprehensive text with models, applications' in *References and DEA-Solver Software*, ed. Kluwer Academic publishers, USA.

Fernandes, E., Capobianco, H. M. P., Ignácio, A. A. V., Sampaio, L. M. D. 2003, `Mercado operador da saúde suplementar', *Fórum de Saúde Suplementar da ANS,* Rio de Janeiro, Brazil. Available at: http://www.ans.gov.br.

Iriart, C., Merhy, E. E., Waitzkin, H. 2001, 'Managed care in Latin America: the new common sense in health policy reform', *Social Science & Medicine* vol. 52, pp. 1243–1253.

Jahanshahloo, G. R., Khodabakhshi, M. 2004, 'Determining assurance interval for non-Archimedean element in the improving outputs model in DEA', *Applied Mathematics and Computation,* vol. 151, pp. 501–506.

Luck, J., Peabody, J. W. 2002a, 'When do developing countries adopt managed care policies and technologies? Part I: policies, experience, and a framework of preconditions', *The American Journal of Managed Care*, vol. 8, no. 11, pp. 997–1007.

Luck, J., Peabody, J. W. 2002a, 'When do developing countries adopt managed care policies and technologies? Part II: policies, experience, and a framework of preconditions', *The American Journal of Managed Care*, vol. 8, no. 12, pp. 1093–1103.

Médici, A. C. 2003, 'Tendências da gestão em saúde ao nível mundial: o caso da Assistência Médica Gerenciada (AMG)', *Fórum de Saúde Suplementar da ANS*, Rio de Janeiro, Brazil. Available at: http://www.ans.gov.br.

Williams, S. J., Torrens, P. R. 1999, *Introduction to health services*, 5th ed. Delmar Publishers. New York, NY.

PRODUCTION FUNCTIONS AND EFFICIENCY EVALUATION IN HEALTH CARE

MARTIN DLOUHY
Faculty of Informatics and Statistics, University of Economics, Prague, Czech Republic.

Abstract

Efficiency evaluation is now widely used in health care at the micro- level as well as macro-level, in the public as well as private sector. In this paper, two major modelling approaches to efficiency evaluation in health care are introduced and critically evaluated. The first group comprises nonparametric methods based on the theory of mathematical programming, such as Data Envelopment Analysis (DEA) and Free Disposable Hull (FDH). The second group originates from econometric analysis of production function. Both groups of modelling techniques are widely used in health care.

Keywords: production function; data envelopment analysis; econometrics; stochastic frontier analysis.

Introduction

Health care is an application area with very specific context and characteristics. The essential situations in health care are quite complex: final output (health) is hard to measure and hard to express in monetary values; the causality between input and output is not always certain; apart from the efficiency of the system, the equity and moral principles have to be borne in mind; there is usually a high variation in demand for services, and so on. But this complexity is the reason why health care as an application area is so challenging for researchers.

We know that to produce health services, doctors, nurses and other staff, buildings, technology, and many other inputs (factors of production) are needed. We also know that there is a relationship between the quantities and combination of the inputs and the level of output. This technological relationship (prices and costs are not involved) is in economic theory described by a production function, which is an abstract model of a real-world production process. This simplified model of production enables economists to study main characteristics of a production process without going into technical details. The production function describes, with help of mathematics, a technical relationship between the inputs and the output, which helps us to understand the nature of production process. The production function can be written as

$$y = f(x_1, x_2, \ldots, x_k), \tag{1}$$

where y is the level of output (total product), and x_j is the quantity of jth input (factor, resource) used in the production. The production frontier represents the maximum amounts of output that can be produced by given amounts of input (in the output maximization model) or, alternatively, the minimum amounts of input required to produce the given amount of output (in the input minimization model). No production unit can achieve efficiency higher than 100%. A technically efficient unit lies on the production frontier.

Managers often measure the level of productivity by the average product, which is the output (total product) per unit of input. However, the right specification of a production function and the knowledge of its parameters offer more than a simple calculation of an average product. There are other types of quantitative economic analysis related to the production function that we are able to quantify:

- the marginal productivities of inputs (the marginal product is the extra output that can be produced by one more unit of a particular input);
- the substitution between inputs and between outputs;
- the returns to scale (the rate at which output increases when all inputs are increased proportionately);
- the technical efficiency of production;
- the technical progress in time (a change in the process of production).

As my prime interest is in efficiency evaluation (i.e.technical efficiency of production), I will not discuss the other types of economic analysis that can be found in practically every textbook of economics (e.g., Samuelson and Nordhaus, 1995, Varian, 1999, Nicholson, 2002), or in the textbook of health economics (Folland, Goodman and Stano, 2004, Feldstein, 1998). There are two major groups of modelling techniques of production function. The first group comprises nonparametric methods based on the theory of mathematical programming, such as Data Envelopment Analysis (DEA) and Free Disposable Hull (FDH). The second group originates from econometric analysis of production function. Both groups of modelling techniques are widely used in health care.

Nonparametric Methods (DEA and FDH)

DEA evaluates efficiency with the help of mathematical programming. Farrell (1957), in his paper on the measurement of efficiency of production, promoted the ideas to specify the production frontier as the most pessimistic piecewise linear envelopment of the data and to construct efficiency measures based on radial uniform contractions or expansions from inefficient observations to the frontier. At that time, Farrell was not able to solve his model for multiple inputs and outputs. The success story of DEA began with the publication of the highly influential paper by Charnes, Cooper, and Rhodes in 1978. Since then a great variety of DEA models with various extensions and modifications have been developed (Charnes, Cooper, Lewin and Seiford, 1994, Cooper, Seiford and Tone, 2002). The journals that published most DEA papers are *European Journal of Operational Research*, *Journal of Productivity Analysis*, *Journal of the Operational Research Society*, and *Management Science* (Tavares, 2002). During two decades DEA has become a very popular method. Tavares (2002) registered in his DEA database 3,230 publications, which had been written by 2,152 authors from 49 countries. Many applications were from health care because the method is attractive to health economists due to its ability to deal with multiple inputs and outputs.

DEA uses quantities of inputs consumed and outputs produced to calculate the relative technical efficiency of a decision-making unit. The relative technical efficiency of the unit is defined as the non-monetary ratio of its total weighted output to its total weighted input or, vice versa, as the ratio of its total weighted input to its total weighted output. DEA allows each unit to choose its own weights of inputs and outputs in order to maximize its efficiency score. For each production unit, DEA (a) calculates the efficiency score; (b) determines the relative weights of inputs and outputs; and (c) identifies peers for each unit that is not technically efficient. The peers of an inefficient unit are technically efficient units with similar combinations of inputs and outputs. The peers serve as benchmarks, which show potential improvements that the inefficient unit can attain. A calculation of the efficiency score by DEA is shown graphically in Figure 1. Let us suppose that there are three production units *A*, *B*, and *C*. These production units produce the same level of single output with two inputs – workers and machines. Furthermore, let us assume that the objective is input minimization. The production units *A* and *B* are technically efficient and they both lie on the production frontier. The production unit *C* uses more inputs than technically necessary; therefore, this unit lies above the production frontier and is technically inefficient. The units *A* and *B* are the peers showing unit *C* how to reduce both inputs to be technically efficient. A hypothetical unit *C** is a linear combination of units *A* and *B*. However, notice that unit *C* can achieve technical efficiency by moving to any position on the production frontier. The hypothetical unit *C** shows the alternative in which all inputs are reduced proportionately.

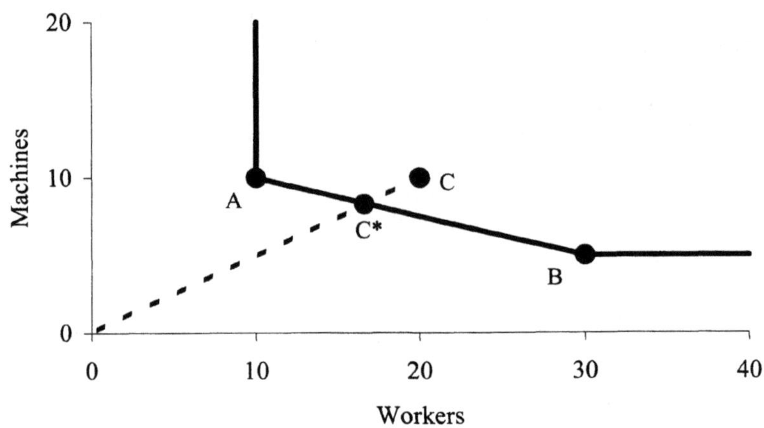

Figure 1: Graphical representation of technical efficiency in DEA

Now let us continue with a mathematical formulation of the DEA model. The mathematical formulation of the so-called *CCR model* (Charnes, Cooper and Rhodes, 1978), which assumes the *constant returns to scale*, is:

$$\text{maximize} \quad \frac{\sum_{i=1}^{r} u_i y_{iq}}{\sum_{j=1}^{m} v_j x_{jq}},$$

$$\text{subject to} \quad \frac{\sum_{i=1}^{r} u_i y_{ik}}{\sum_{j=1}^{m} v_j x_{jk}} \leq 1, \quad k = 1, 2, \ldots, n, \tag{2}$$

$$u_i \geq \varepsilon, i = 1, 2, \ldots, r,$$
$$v_j \geq \varepsilon, j = 1, 2, \ldots, m,$$

where x_{jk} is the amount of input j used by unit k, y_{ik} is the amount of ith output produced by unit k, weights u_i and v_j are variables in the model, ε is infinitesimal constant. The value of objective function is the efficiency score of unit q. This program (2) is an input-oriented version of the CCR model. To solve the program (2) by the linear programming, the program has to be transformed to the linear form

$$\begin{aligned} \text{maximize} \quad & \sum_{i=1}^{r} u_i y_{iq}, \\ \text{subject to} \quad & \sum_{i=1}^{r} u_i y_{ik} \leq \sum_{j=1}^{m} v_j x_{jk}, \quad k = 1, 2, \ldots, n, \\ & \sum_{j=1}^{m} v_j x_{jq} = 1, \\ & u_i \geq \varepsilon, i = 1, 2, \ldots, r, \\ & v_j \geq \varepsilon, j = 1, 2, \ldots, m. \end{aligned} \tag{3}$$

The model (3) has to be formulated and solved for each unit q=1, 2, ..., n. For the computation, it is useful to consider a dual model. The dual model involves fewer constraints than the primal model, and therefore, the dual model is the preferred model to solve. The formulation of the dual model in the matrix form is:

$$\begin{aligned} \text{minimize} \quad & \theta_q - \varepsilon(\mathbf{e}^T\mathbf{s}^+ + \mathbf{e}^T\mathbf{s}^-), \\ \text{subject to} \quad & \mathbf{X}\boldsymbol{\lambda} + \mathbf{s}^- = \theta_q \mathbf{x_q}, \\ & \mathbf{Y}\boldsymbol{\lambda} - \mathbf{s}^+ = \mathbf{y_q}, \\ & \boldsymbol{\lambda} \geq \mathbf{0}, \mathbf{s}^+ \geq \mathbf{0}, \mathbf{s}^- \geq \mathbf{0}, \end{aligned} \tag{4}$$

where θ_q is the efficiency score, $\boldsymbol{\lambda}$ is the vector of variables, $\mathbf{s}^+$, $\mathbf{s}^-$ are vectors of slack variables, $\boldsymbol{x_q}$ and $\boldsymbol{y_q}$ are vectors of inputs and outputs of the evaluated unit, $\mathbf{X}$ is a matrix of m inputs and $\mathbf{Y}$ is a matrix of r outputs, $\mathbf{e}^T = (1, 1, \ldots, 1)$, and ε is infinitesimal constant. The production unit q is technically efficient if the optimal value of variable θ^*_q is one and all slack variables s_i^{+*} and s_i^{-*} equal zero (the symbol "*" denotes the optimal value). The output-oriented model version of the CCR model is written as

$$\begin{aligned} &\text{maximize} && \phi_q - \varepsilon(\mathbf{e}^T\mathbf{s}^+ + \mathbf{e}^T\mathbf{s}^-), \\ &\text{subject to} && \mathbf{X\lambda} + \mathbf{s}^- = \mathbf{x_q}, \\ &&& \mathbf{Y\lambda} - \mathbf{s}^+ = \phi_q\mathbf{y_q}, \\ &&& \boldsymbol{\lambda} \geq \mathbf{0}, \mathbf{s}^+ \geq \mathbf{0}, \mathbf{s}^- \geq \mathbf{0}, \end{aligned} \tag{5}$$

where ϕ_q is the efficiency score. The meaning of all other parameters and variables is the same as in the input-oriented model. A production unit q is technically efficient if the optimal value of ϕ^*_q is one and all s_i^{+*} and s_i^{-*} equal zero.

According to the returns to scale, the theory distinguishes four alternative DEA models: the constant-returns-to-scale model, the variable-returns-to-scale model, the non-increasing-returns-to-scale-model, and the non-decreasing-returns-to-scale model. These models differ only in restriction for the vector $\boldsymbol{\lambda}$ (see Table 1).

Returns to scale	**Formulation**
Constant returns to scale (the CCR model)	$\boldsymbol{\lambda} \geq \mathbf{0}$
Variable returns to scale	$\boldsymbol{\lambda} \geq \mathbf{0}, \mathbf{e}^T\boldsymbol{\lambda} = 1$
Non-increasing returns to scale	$\boldsymbol{\lambda} \geq \mathbf{0}, \mathbf{e}^T\boldsymbol{\lambda} \leq 1$
Non-decreasing returns to scale	$\boldsymbol{\lambda} \geq \mathbf{0}, \mathbf{e}^T\boldsymbol{\lambda} \geq 1$

Table 1: Returns to scale in DEA models

Free Disposable Hull (FDH) was formulated by Deprins, Simar, and Tulkens in 1984. Advocates of FDH argue that there are no a priori economic grounds to believe that the production frontier is convex or non-convex, and that this flexibility is the advantage of FDH, which does not require any a priori assumption about the returns to scale. Let us have n units, which consumes m inputs and produce r outputs. A mathematical formulation of the input-oriented FDH model follows:

$$\begin{aligned} &\text{minimize} && \theta_q - \varepsilon\,(\mathbf{e}^T\mathbf{s}^+ + \mathbf{e}^T\mathbf{s}^-), \\ &\text{subject to} && \mathbf{X\lambda} + \mathbf{s}^- = \theta\,\mathbf{x_q}, \\ &&& \mathbf{Y\lambda} - \mathbf{s}^+ = \mathbf{y_q}, \\ &&& \mathbf{e}^T\boldsymbol{\lambda} = \mathbf{1}, \lambda_k \text{ - binary}, \mathbf{s}^+ \geq \mathbf{0}, \mathbf{s}^- \geq \mathbf{0}, \end{aligned} \tag{6}$$

where $\mathbf{Y}$ is the matrix of outputs (r, n), $\mathbf{X}$ is the matrix of inputs (m, n), $\boldsymbol{\lambda}$ is the vector of variables, $\mathbf{s}^+$ a $\mathbf{s}^-$ are the vectors of slack variables, $\mathbf{e}^T = (1,1,...,1)$, and ε is infinitesimal constant. The program (6) can be solved by mixed binary programming. The value of objective function is the technical efficiency. Generally, the FDH efficiency scores are higher than the DEA scores, because a unit cannot be dominated by a convex combination of other units in the FDH model. Notice that the FDH model differs from convex DEA models only with the constraint for the vector λ. FDH can be used for evaluation of technical efficiency in the health sector like DEA. Nevertheless, applications of data envelopment analysis dominate the literature (Hollingsworth, 2003), whereas FDH is used only exceptionally.

Econometric Analysis

Econometric analysis, or econometrics, can be described as a scientific discipline that is based on an amalgam of economic theory, mathematical economics, economic statistics, and mathematical statistics (Gujarati, 2003). The foundations of econometric analysis can be found in many econometrics textbooks (e.g., Gujarati, 2003, Ruud, 2000, Maddala, 2001), I will, therefore, introduce the foundations of econometric analysis very briefly.

A main tool of econometrics is the regression analysis that studies the dependence of one variable, called the *dependent variable*, on one or more other variables, called the *independent* or *explanatory variables*. The method of *ordinary least squares* (OLS) is the most frequently used technique of the regression analysis. A magic of the ordinary least squares stems from the fact that the least-square estimates posses some optimal properties for which the least-square estimates are the best estimators. An econometric model is formulated in the form of one or more equations. To begin, you usually start with a linear relationship between the dependent and the explanatory variables. A linear regression function is written as

$$y = \beta_0 + \beta_1 x_1 + \beta_2 x_2 + \beta_k x_k + u, \qquad (7)$$

where y is the dependent variable, $x_1, x_2, \ldots, x_k$ are the explanatory variables, β_0, $\beta_1, \ldots, \beta_k$ are the regression coefficients, and u is the stochastic error term. The stochastic error term represents random shocks, but also effects of less the important, unobserved variables. The ordinary least squares can be applied, with appropriate transformations, to some nonlinear models. This is important because the famous Cobb-Douglas production function (Cobb and Douglas, 1928) is also a nonlinear model.

The first step of econometric analysis lies in using economic theory for the formulation of a hypothetical relationship between the dependent and explanatory variables. This step of analysis is called a *specification*. A model is correctly specified if it includes all important variables, and if the right functional form of the model is chosen (linear or nonlinear). If you do not include some important variables in your model, or you choose a wrong functional form, econometricians will say that there is a *specification error* in your model. This is a nice way to say that your econometric model is wrong. Let us now suppose that you succeeded in the specification of your model. Then, the second step of analysis is the estimation of the regression coefficients $\beta_0, \beta_1, \ldots, \beta_k$. The third step of analysis is testing the statistical significance of the individual regression coefficients and of the overall significance of the regression model. This step is called *hypothesis testing*.

The most popular classical production function is the *Cobb-Douglas production function*, which was developed by Cobb and Douglas in 1928. The Cobb-Douglas production function is specified in the form

$$y = aK^{\alpha}L^{\beta}, \qquad (8)$$

where y is the output (total product), K is the capital, and L stands for the labour, and a, α, and β denote the regression coefficients to be estimated. The parameter α is the elasticity of output with respect to capital input, and β is the elasticity of output with respect to labour input. In the original form, it was supposed that $\alpha + \beta = 1$ (Cobb and Douglas, 1928). Generally, the Cobb-Douglas production function, in its stochastic form, can be expressed as

$$y = ax_1^{\alpha_1}x_2^{\alpha_2}\ldots x_k^{\alpha_n}e^{u}, \qquad (9)$$

where y is the output, x_j are the inputs, u is the stochastic error term, and e is the base of natural logarithm. We easily transform this nonlinear relationship between the output and inputs (9) to the linear form

$$ln\, y = ln\, a + \alpha_1\, ln\, x_1 + \alpha_2\, ln\, x_2 + \ldots + \alpha_n\, ln\, x_k + u. \qquad (10)$$

The model (10) is linear in the parameters, so the ordinary least squares can be used for estimating the Cobb-Douglas production function. The interpretation of parameters in the Cobb-Douglas production function (10) is as follows:

a) Parameter a is interpreted as a technical level of production process.
b) Parameter α_j is the partial elasticity of output to input j. This parameter says what percentage change in output is induced by one percentage change in input. If $\alpha_j = 0.5$, then one percent increase of input j will lead to a 0.5% increase in output (ceteris paribus).
c) The sum of parameters α_j measures the relative change in output to a proportionate relative change in inputs, that is, it determines the returns to scale. If this sum equals one, then increasing the quantities of all inputs by one percent will increase the output by one percent. Such production function describes the production process with the constant return to scale. The increasing or decreasing returns to scale are two other alternatives.

Economic theory and empirical econometric analysis developed a variety of production functions. The choice of production function depends on the nature of production process, which is to be modelled, and also the objective of the analysis.

Frontier Analysis

The interpretation of a regression model is simple if one assumes that the residual represents only inefficiency. If there is no recognition of statistical error, the level of efficiency is calculated as a distance between the observation and the regression line. If the residual is positive, the unit is relatively efficient, and vice versa, if the residual is negative, the production unit is relatively inefficient. The shortcoming of such an approach is that it concentrates on the estimation of average behaviour, not on the best performance. In economic theory, a production function is defined as a maximum output obtainable from given resources, not as an average or usual level of output. The classical regression analysis of production function was, therefore, extended to the *deterministic frontier analysis* and *stochastic frontier analysis* (Kumbhakar and Lovell, 2000). To begin, let us assume that production model can be written as

$$y_i = f(x_i;\beta)\cdot TE_i, \tag{11}$$

where y_i is the scalar output of ith unit (producer, hospital), x_i is the vector of n inputs used by ith unit, $f(x_i;\beta)$ is the production frontier, β is the vector of unknown technology parameters to be estimated, and TE_i is the output-oriented technical efficiency of unit i. The output-oriented technical efficiency TE_i is defined as the ratio of observed output y_i to the maximum feasible output, which is determined by the production frontier $f(x_i;\beta)$. We have

$$TE_i = \frac{y_i}{f(x_i;\beta)}. \tag{12}$$

If the observed unit is technically efficient, then technical efficiency $TE_i = 1$; if the unit is not technically efficient, then technical efficiency $TE_i < 1$. This type of the frontier model is called the *deterministic frontier analysis (DFA)* because the entire difference between the observed output y_i and maximum feasible output $f(x_i;\beta)$ is explained by the technical inefficiency. There is no stochastic term, which could represent random shocks in production. In comparison, the *stochastic frontier analysis (SFA)* assumes the existence of random shocks that affect efficiency of the production process. The deterministic model is modified to a stochastic model as

$$y_i = f(x_i;\beta)\cdot \exp\{v_i\}\cdot TE_i, \tag{13}$$

where $[f(x_i;\beta)exp\{v_i\}]$ is the stochastic production frontier, which consists of two parts: the deterministic part $f(x_i;\beta)$ and the stochastic part $exp\{v_i\}$, where v_i is the two-sided random component with the normal distribution $N(0,\sigma_v^2)$. The output-oriented technical efficiency is given by the ratio

$$TE_i = \frac{y_i}{f(x_i;\beta)\exp\{v_i\}}. \tag{14}$$

Similarly as in the deterministic frontier model, the production unit is technically efficient if $TE_i = 1$ and inefficient if $TE_i < 1$. Advocates of the stochastic frontier analysis argue that it is not possible to ignore random shocks as the deterministic frontier analysis proposes, and therefore, the stochastic frontier analysis should be preferred.

Deterministic Frontier Analysis

Let us rewrite the frontier model (11) as

$$y_i = f(x_i;\beta)\cdot exp\{-u_i\}, \tag{15}$$

where the technical efficiency TE_i is substituted by $exp\{-u_i\}$. We assume $u_i \geq 0$ to guarantee that the technical efficiency is less than one. In case that the frontier $f(x_i;\beta)$ takes the form of the Cobb-Douglas production function, the deterministic production frontier is expressed as

$$\ln y_i = \ln \beta_0 + \sum_{j=1}^{k} \beta_j \ln x_{ij} - u_i \cdot \qquad (16)$$

One of the possible estimation methods of deterministic frontier analysis is known as the *Corrected Ordinary Least Squares (COLS)*. This method, suggested by Winsten (1957), estimates production frontier in two steps. In the first step, the ordinary least squares method is applied to obtain consistent and unbiased estimates of the slope parameters β_j and consistent but biased estimate of intercept β_0. In the second step, the intercept is "corrected" by adding a residual with maximal value

$$\hat{\beta}_0^* = \hat{\beta}_0 + \max_i\{\hat{u}_i\}. \qquad (17)$$

The production unit with a maximum value of residual $\hat{u}_i$ is from the definition automatically efficient and lies on the production frontier. Residuals of OLS regression are corrected by

$$-\hat{u}_i^* = \hat{u}_i - \max_i\{\hat{u}_i\}. \qquad (18)$$

The residuals of COLS are non-negative, with at least one being zero (the efficient unit). The technical efficiency of the production unit is calculated by $TE_i = exp\{-\hat{u}_i^*\}$. The main advantage of COLS is its simplicity; the main disadvantages of the method are ignoring the stochastic term and assuming that the efficient production frontier is parallel to the OLS regression (Kumbhakar and Lovell, 2000).

Stochastic Frontier Analysis

The original papers on the stochastic frontier analysis were simultaneously published by two research teams in 1977 (Meeusen and van den Broeck, 1977; Aigner, Lovell and Schmidt, 1977). Methods of the stochastic frontier analysis assume that a difference between the observed and the feasible output can be partitioned into two components, one representing technical inefficiency and one representing effects of random shocks. If we, again, assume the case of the Cobb-Douglas production function then we can write the stochastic frontier analysis model as

$$\ln y_i = \ln \beta_0 + \sum_{j=1}^{k} \beta_j \ln x_{ij} + v_i - u_i \text{ ,} \tag{19}$$

where v_i is the two-sided random component with the normal distribution $N(0, \sigma_v^2)$ and u_i is the non-negative technical efficiency component of error term with the non-negative half normal distribution $N(0, \sigma_u^2)$. The error components u_i and v_i are independent of each other, and of the regressors. The composed error term ε_i is defined as $v_i - u_i$. The composed error term is asymmetric since $u_i \geq 0$. Estimation of technical efficiency u_i relies on the assumption about its distribution. The distribution of u_i is usually assumed to be non-negative half normal, truncated normal, exponential or gamma (Kumbhakar and Lovell, 2000). According to distributions of v_i and u_i, the frontier models are called as the normal-half normal model, the normal-exponential model, the normal-truncated normal model, and the normal-gamma model. The restrictive assumptions about the distribution of u_i may be considered as a weakness of the stochastic frontier analysis. I am not going to present all advanced estimation methods used in the stochastic frontier analysis. I will show calculations for the corrected ordinary least squares and for two versions of the normal-half normal model on the illustrative example below. These estimation methods are now incorporated in statistical and econometric software.

I continue with an illustrative example. Figure 2 shows 17 hospitals from the Region of West Bohemia, the Czech Republic (the data comes from Dlouhy and Pankova, 2003). The single output is the number of discharges; the single input is the number of physicians in full-time equivalents. The Cobb-Douglas production function was estimated by three methods: by the ordinary least squares, by the deterministic production frontier after the "correction" by the corrected ordinary least squares (COLS), and by the stochastic production frontier. You can see that the deterministic frontier estimated by COLS is the most optimistic production frontier. This means that COLS finds higher inefficiencies than stochastic frontier methods and ordinary least squares regression.

Applications

Both groups of efficiency analysis of production are now widely used in health care at the micro-level as well as macro-level, in the public and private sector, and in developed as well as developing countries. As an illustration, we present a few examples.

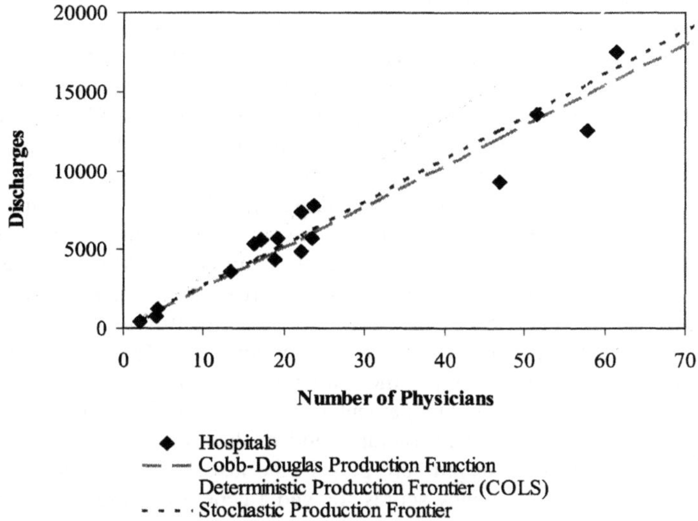

Figure 2: Production frontiers

Ozgen and Ozcan (2002) examined market competition and facility characteristics that can be related to technical efficiency in the production of multiple dialysis outputs. They used the 1997 U.S. data on 791 Medicare-certified free-standing dialysis facilities. In their model, they included three outputs and nine inputs. Technical efficiency scores were estimated by an input-oriented variable-returns-to-scale DEA model. A binary variable of efficiency status was then regressed (multivariate logistic regression analysis) against market and facility characteristics and control factors. A significant association was found between efficiency and type of ownership, with non-profit and government-owned facilities more likely to be inefficient producers. On the other hand, the affiliation with large-sized chains, as compared to independent dialysis facilities, was associated with decreasing efficiency. This suggests that growing consolidation of the dialysis industry into large facility chains may not be a strategy for gaining better technical efficiency. A limitation of the study is the absence of measures to adjust for variability in case mix across facilities. Since the study examined a homogenous study sample of facilities that produce basically one type of output (dialysis treatments) to patients with the same disease (chronic renal failure), incorporation of case-mix variables in the efficiency model may be considered less necessary.

Ozcan, Merwin, Lee, and Morrissey (2004) analysed technical efficiency of community mental health centres (CMHCs). They examined the longitudinal patterns of efficiency over a five-year period (1994–1998), before and after implementation of the mandatory Medicaid managed care plan. By an input-oriented variable-returns-to-scale DEA model, they compared efficiency scores between the managed care site – the Tidewater area – and the control site – Richmond, Virginia. Medicaid managed care for behavioural care was implemented in 1996 at the Tidewater area. Output variables were the number of Medicaid seriously mentally ill patients in the Medicaid eligibility category of supplemental security income (SSI), and the number of Medicaid patients except SSI. This categorization of output was a proxy for outputs "more severe "and "less severe", and hence served as the case-mix difference for outputs. The inputs were the non-emergency crisis support in CMHC, the outpatient assessment, the outpatient therapy, the outpatient medication management, the clubhouse services, and case management services. In the Tidewater area, three services – outpatient assessment, therapy, and medication management – were covered by a capitated managed care plan in the last three years of the study. The five-year trend of efficiency scores showed a generally increasing trend for the Tidewater CMHCs, but stagnation, even retrenchment in the Richmond CMHCs.

Dlouhy and Pankova (2003) investigated a hospital performance over a period of four years. We sought to identify potential inefficiencies and overall trends in efficiency of hospitals in the region of West Bohemia, the Czech Republic, 1996–1999. The efficiency of hospitals was evaluated by an input-oriented variable-returns-to-scale model. The hospital performance decreased during the observed period with a possible reversal in the last year. A change in the reimbursement mechanism was suggested as one of the possible explanations of that trend. The introduction of strict budgeting in July 1997, which replaced less regulated fee-for-service system, resulted in a significant impact on the behaviour of hospitals. Trends for individual hospitals, however, showed large variations that could not be easily interpreted.

I discussed (Dlouhy, 2001) DEA as an alternative method of a health-system performance evaluation. My work was inspired by the *World Health Report 2000* (World Health Organization, 2000). The *Report* identified five objectives of the health system whose relative weights were obtained through a survey of 1,006 respondents. In the *Report*, resources were measured by per capita health expenditures in international dollars. The data on 191 countries came from the statistical annex of the *World Health Report 2000* and were analysed by the output-oriented constant-returns-to-scale model with five outputs and one input (per capita health expenditures in international dollars).

Hollingsworth (2003) reviewed 188 published papers on the frontier efficiency measurement in health care. The review mainly included DEA-based methods that dominated the literature, but there were an increasing number of parametric techniques, such as stochastic frontier analysis. DEA alone was used in 50% of studies; a quarter of studies used regression analysis in two stage analysis, typically to regress factors on the efficiency scores in an attempt to determine influences on efficiency. Stochastic frontier analysis and other parametric frontier techniques were used in 12% of studies; Malmquist techniques were used in 9% of studies.

Conclusion

Efficiency evaluation is now widely used in health care at the micro-level as well as macro-level, in the public as well as private sector, and in developed as well as developing countries. In this paper, I have described and evaluated two major approaches to efficiency evaluation in health care: nonparametric methods, such as DEA and FDH, and econometric analysis of production function.

In my view, the strengths of DEA are: a) the method is able to deal with multiple inputs and outputs; b) for technically inefficient units, DEA identifies the peers that are real production units, so the efficiency improvements should be attainable; c) software for DEA is now available, which makes it easy to carry out all calculations and present results. The weaknesses are: a) a deterministic nature of method. Because DEA does not include an error term in the model, there is a possibility of a measurement error; b) hypothesis testing is much less developed in comparison to econometric analysis.

The strengths of FDH are: a) the method is able to deal with multiple inputs and outputs; b) in contrast to DEA, no assumptions about the nature of returns to scale are necessary. The weaknesses are: a) deterministic nature of the method; b) hypothesis testing is much less developed in comparison to econometric analysis; c) for small samples, FDH is less powerful in discriminating between the units in comparison to DEA.

In my experience the main advantage of econometric analysis over non-parametric methods, such as DEA and FDH, is a highly developed methodology of hypothesis testing. Both the selection of explanatory variables and the selection of functional form can be tested. The weaknesses are: a) it is more complicated to deal with multiple inputs and outputs in econometric analysis than in DEA and FDH; b) it is necessary to cope with many problems related to the estimation method and the nature of data such as the multicollinearity, autocorrelation, heteroscedasticity etc. Stochastic Frontier Analysis is a special econometric method that incorporates random shocks in efficiency evaluation. The weakness of this method is that assumptions about the distribution of technical efficiency have to be made.

Acknowledgements

This work was supported by the project no. 402/03/1360, funded by the Grant Agency of the Czech Republic.

References

Aigner DJ, Lovell CAK and Schmidt P (1977). Formulation and estimation of stochastic frontier production function models. *Journal of Econometrics* 6: 21–37.

Charnes A, Cooper WW, Lewin, AY and Seiford LM (eds) (1994). *Data envelopment analysis: theory, methodology and applications.* Kluwer Academic Publishers: Boston.

Charnes A, Cooper WW and Rhodes E (1978). Measuring the inefficiency of decision making units. *Eur J Opl Res* 2: 429–444.

Cobb CW and Douglas PH (1928). A theory of production. *American Economic Review*, 18: 139–165.

Cooper WW, Seiford L W and Tone K (2000). *Data envelopment analysis.* Kluwer Academic Publishers: Boston.

Deprins D, Simar L and Tulkens H (1984). Measuring labour efficiency in post offices. In Marchand M, Petieau P and Tulkens H (eds). *The performance of public enterprises: concepts and measurement.* North Holland: Amsterdam.

Dlouhy M (2001). An alternative of health system performance measurement in the context of World Health Report 2000. *The Seventh International Conference on Theoretical and Practical Aspects of Public Finance 2001.* University of Economics Prague: Prague.

Dlouhy M and Pankova V (2003). Hospital performance and trends. In Rauner MS and Heidenberger K (eds). *Quantitative Approaches in Health Care Management.* Peter Lang: Frankfurt am Main.

Feldstein PJ (1998). *Health care economics*. Fifth Edition, Delmar Publishers.

Folland S, Goodman AC and Stano M (2004). *The economics of health and health care*. Fourth Edition, Prentice Hall: Upper Saddle River.

Gujarati DN (2003). *Basic econometrics*. Fourth Edition, McGraw-Hill: New York.

Hollingsworth B (2003). Non-parametric and parametric applications measuring efficiency in health care. *Health Care Management Science* 6: 203–218.

Kumbhakar SC and Lovell CAK (2000). *Stochastic frontier analysis*. Cambridge University Press: Cambridge.

Maddala GS (2001). *Introduction to econometrics*. Third Edition, John Wiley & Sons: Chichester.

Meeusen W and van den Broeck J (1977). Efficiency estimation from Cobb-Douglas production functions with composed error. *International Economic Review* 18: 435–444.

Nicholson W (2002). *Microeconomic theory: basic principles and extensions*. Eighth Edition, South-Western.

Ozcan YA Merwin E, Lee K and Morrissey JP (2004). Benchmarking using DEA: the case of mental health organizations. In Brandeau ML, Sainfort F and Pierskalla WP (eds). *Operations research and health care: a handbook of methods and applications*. Kluwer Academic Publishers: Norwell.

Ozgen H and Ozcan YA (2002). A national study of efficiency for dialysis centres: An examination of market competition and facility characteristics for production of multiple dialysis outputs. *Health Services Research* **37**: 711–732.

Ruud PA (2000). *An introduction to classical econometric theory*. Oxford University Press: New York and Oxford.

Samuelson PA and Nordhaus WA (1995). *Economics*. 15th Edition, McGraw-Hill: New York.

Tavares G (2002). *A bibliography of data envelopment analysis (1978–2001)*. RUTCOR Research Report 01-02, Rutgers University: Piscataway, NJ.

Varian HR (1999). *Intermediate microeconomics: a modern approach*. Fifth Edition, W. W. Norton & Company: New York and London.

Winsten CB (1957). Discussion on Mr. Farell's Paper. *Journal of the Royal Statistical Society*, Series A, General, 120, Part 3, 282–284.

World Health Organization (2000). *The world health report 2000 – health systems: improving performance*. Geneva.

PATIENT ISSUES

CHOICE *OR* CHOICE?

PENELOPE M. MULLEN
University of Birmingham, UK

Abstract

Freedom and autonomy are critical to our well-being, and choice is critical to freedom and autonomy. The provision of greater choice is an increasing theme in many healthcare systems. Choice in healthcare arises at many different levels. In the US choice of health plan – a choice which is being introduced in the Netherlands – has long been important. Patient choice in care and treatment is increasingly expected. In England, the 'Choice Agenda' involves greater choice of hospital for elective care. But what is or should be OR's role in relation to choice in healthcare?

From the patient's perspective, OR has much to offer. For example, provision of appropriate information, decision analysis and multi-criteria decision-making approaches could assist in making choices. OR could help institutions wishing to attract or deter patients, to forecast and to channel choices. For healthcare systems and healthcare planners, OR can make a very important contribution in modelling actual and potential patient choices of health plan, institutional provider or type of treatment, both to determine the consequences and to assist in planning to accommodate those choices.

However, such contributions would appear to treat choice as a neutral concept to be accommodated by technical approaches. But, as Schwartz points out, choosing well is difficult. Further, there is increasing theoretical and empirical literature reporting the negative consequences of choice; to the individual (for example, anxiety and regret); to the healthcare system (for example, de-stabilising of provision); and to society (for example, increased inequality and reduced health levels). However, suggestions that choice should be limited could be construed as patronising and a denial of freedom and autonomy.

Should OR consider choice as a neutral concept and offer its technical assistance to the different parties? Or should OR seek to address the negative consequences and attempt to contribute to the exposure, understanding and, if possible, alleviation of them? If the latter, how? What is or could be the contribution of OR? This paper explores these issues against the background of political developments in the UK and elsewhere.

Keywords: health; health care; choice; Operational Research.

Introduction

As a principle, choice is a good thing. It is difficult, in the abstract, to argue against choice and it could appear patronising to suggest that choice should be restricted or denied. According to Schwartz (2004, p.99) "Freedom and autonomy are critical to our well-being, and choice is critical to freedom and autonomy." However, choosing is difficult and an increasing number of commentators and studies are pointing to dilemmas associated with choice.

After identifying areas of choice in healthcare, this paper outlines policies relating to the English NHS 'Choice Agenda'. After listing some of the many claimed advantages and disadvantages of these policies, the paper then turns to examining the potential contribution of OR from a number of perspectives. However, the paper then moves on to the question of whether choice is 'a good thing', exploring a number of difficulties and adverse effects. In the light of these, and proposals for their mitigation, the paper returns to the subject of the potential contribution of OR.

The distinction between individual choice and collective choice, the latter often using democratic and/or other participatory approaches, needs to be noted. In many aspects of health and healthcare, collective or community choices can be far more important, both for the participants and community, and for improved health outcomes, than individual choice. Further, there can be considerable tensions and even conflict between increasing individual choice and increasing collective choice. In the light of current policies, this paper focuses on individual choice, but that should in no way be construed as disregarding the importance of collective and community participation and choice in health care.

Choice in Healthcare

Individual choice can occur in many different aspects of healthcare. A simple search on Choice + Health yielded over 30,000 hits. Choices can relate to timing and manner of treatment, for example, choice between alternative treatments, choice in timing of treatment, choice of doctor or indeed choice of which type of professional to approach and be treated by, and choice of hospital or other healthcare provider. Rather more fundamentally, there can be choice of Health Plan which, whilst a common choice in the USA, is being increasingly introduced in Europe, for example in Germany and the Netherlands. Choices between Plans can involve whether or not they permit, for example, choice of doctor and hospital, but also can involve financial choices, for example the level of premiums versus risk borne by the consumer, the existence and levels of co-payments and deductibles, as well as overall coverage.

Other choices in healthcare include the right to refuse treatment, the choice of a 'good death' (Cook, 2003) and choice of time of death. There can be choice of whether or not to be involved in trials and in teaching of students, whether or not to participate in screening or to be, or to allow one's children to be, vaccinated or immunised, in addition to choices about lifestyle. On another dimension, Charnock (2004) argues that patients should be able to choose to have access to email consultation.

English[1] 'Choice Agenda'

In many countries, as Dawson et al (2004, p.195) note, "patient choice is a routine part of the normal healthcare system..." However, the main, though not exclusive, focus of this paper is the current English NHS 'Choice Agenda'. Although there are a number of aspects of this, much current interest is focused on choice of provider (hospital) for elective, scheduled, acute care.

Associated with this policy is a major change in the way that secondary-care providers (hospitals) in England are funded, with the introduction of a policy termed 'Payment by Results' (PbR) (DoH 2002). Under this policy, which is effectively cost-per-case reimbursement, payment to hospitals will be "linked to activity and adjusted for casemix" (DoH 2005b) and "hospitals (and other providers) will be paid for the activity that they undertake...on the basis of a standard national price tariff" (DoH 2002, p.8). As outlined in Figure 1, funds for healthcare will continue to be passed, on a weighted capitation basis, to Primary Care Trusts (PCTs) which have responsibility for providing or securing healthcare for their populations, currently averaging 150,000 – a figure which is likely to increase in the near future as a result of mergers (DoH, 2005c). However, unlike previously, hospitals will be funded entirely on a cost-per-case

[1] Although the UK has a single National Health Service (NHS), there are increasing differences between the four countries in the UK in its organisation and delivery. Thus, reference here to England means England alone and not the rest of the UK.

basis with no core funding. Such PbR will, according to DoH (2004, p70), "support patient choice by enabling funds to follow individual patient choices for treatment".

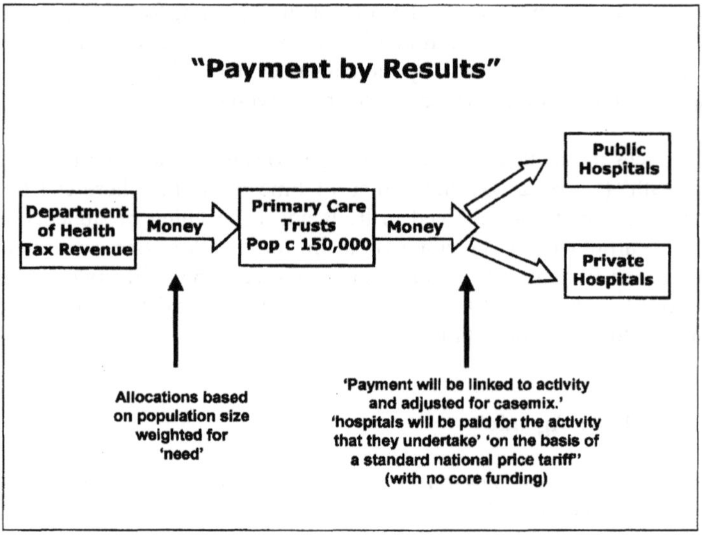

Figure 1: Funding Flows in the English NHS

In examining the policy on choice of hospital, it is necessary to note that in the UK NHS, primary-care doctors (GPs) act as gatekeepers to secondary care and so virtually all access to elective secondary care is via referral. The choice policy, as outlined in DoH (2004, para 2.10 p.28) means by "By December 2005 all patients who need surgery will be offered a choice of four to five alternatives at the time they are referred for treatment by their GP". To secure that, "PCTs will be responsible for making sure that from 2006 they offer... a choice of four or five local NHS providers, together with all NHS Foundation Trusts [NHS hospitals with greater independence as Public Benefit Corporations] and nationally procured Independent Sector [non-NHS] Treatment Centres" (DoH, 2005a, p.5). It is intended to extend this policy so that "By 2008, patients referred by their GP will be able to choose any provider [public or private] able to meet NHS standards and to deliver care at tariff." (DoH 2004, p.70)

Whilst such proposals, especially in respect of choice, will appear unexceptional to many from different healthcare systems, they have generated considerable debate in England. The government argues that they will improve quality, increase responsiveness, reduce health inequalities by giving choice to all social classes, reduce waiting times, reduce costs and increase efficiency (DoH, 2004; Lewis, 2005). The increased competition between hospitals and increased involvement of the private sector will, it is claimed, contribute to these benefits. Somewhat more cynically, some argue that there is a deliberate attempt to destabilize the system in the belief that this will result in improved performance. From the other side, however, it is argued that PbR and patient choice will lead to hospitals and other providers 'cream skimming' the easier cases (Coombes, 2005), have an adverse effect on health outcomes (Appleby and Dixon, 2004), increase health inequality, restrict and lead to greater inequity in access to services (Deeming and Keen, 2004), increase costs and lead to financial crises for hospitals (Cole 2005), require excess capacity, fragment services and reduce continuity of care. In addition, there are concerns about increased GP workload (Iles, 2003) and, more simply, that the necessary IT systems do not work (de Kare-Silver 2005). There are also claims that the policies do not address the main issues of concern to patients and to healthcare. Rosen et al (2005, p.6) reported that a common opinion expressed in a series of focus groups was "I don't actually want choice. All I want is good quality".

Each of these, and the many other claimed advantages and disadvantages, is interesting and could warrant a paper in its own right. However, the primary focus here is on the potential contribution of Operational Research to choice in healthcare, in general, and the English 'choice agenda' in particular.

Possible OR Contribution

OR's contribution could be viewed from a number of different perspectives: the patient's perspective; the hospital's perspective and the perspective of the healthcare system as a whole – policy makers and planners. Again, almost any one of these areas of potential contribution could warrant a paper in its own right.

The Patient's perspective: OR could assist patients in making choices by the provision of appropriate information and in eliciting preferences (Mullen and Spurgeon, 2000). Patients could be assisted in using a range of multi-criteria decision-making approaches, for example VISA (see Figure 2) and, where risk or uncertainly are involved, decision analytical approaches might be useful.

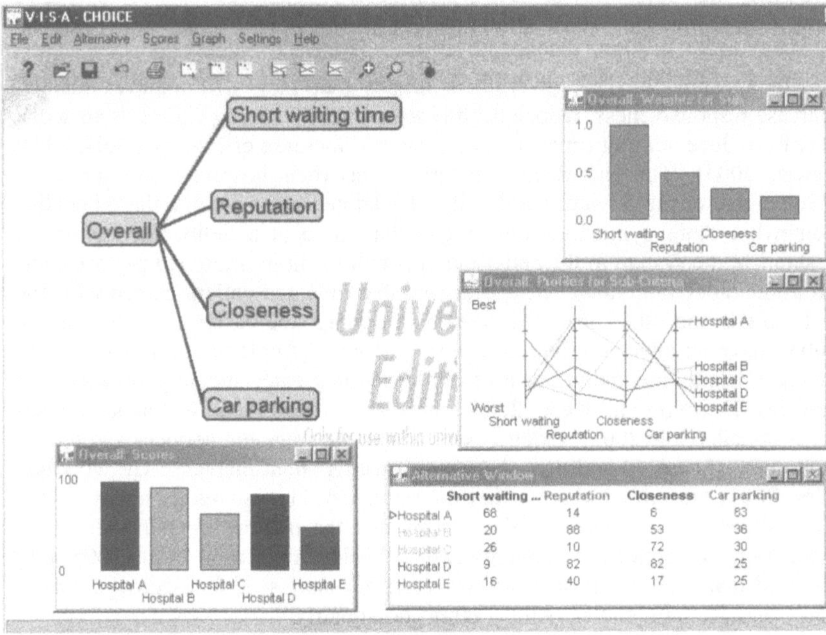

Figure 2: Assisting in choice of hospital using VISA

The Hospital's Perspective: Possible contributions could include determining the optimum casemix to maximise revenue or profit and assisting in the understanding of patient preferences. OR techniques might be used to assist profit-maximising hospitals in channelling patient choice, both to attract and to deter patients. Forecasting and modelling the effects of choice could make a contribution, as could capacity planning.

The Healthcare System's Perspective: Policy Makers and Planners Contributions here could include modelling the potential for choice of provider taking account of access/distance, and mapping constraints on choice (Damiani et al 2005). OR could assist in the understanding of patient preferences at a systems level, and in forecasting and modelling various aspects of actual and potential patient choice. Capacity planning to accommodate these choices and deal with the consequences of choice could also contribute. OR could also help in determining the system-wide consequences in terms of finance, equity and access.

But is Choice a Good Thing?

Until this point, it has been assumed that 'choice' is axiomatically 'a good thing'. Clearly there are problems with choice, including limits to choice and resistance to choice.

Limits to Choice: There are physical and legal limits to choice. Some treatments and modes of care are unlawful in all or some countries. Compulsory quarantine restricts the choice of individuals, as many found in the 2003 SARS crisis. Lack of competence can limit choice, relatively non-controversially in the case of mentally non-competent adults and young children, but some cases are less clear-cut. In an interesting discussion on balancing rights and responsibilities in adolescent care, Larcher (2005, p.354) concludes that "an adolescent's right to refuse treatment depends on the circumstances". Choice for some limits choice for others – the 'best' hospitals and doctors have finite capacity. It is also questioned whether there should be limits to choice of less (cost) effective treatment in collectively or publicly-funded healthcare systems (King's Fund, 2003; Sharpe, 2002)

Resistance to Choice: There appears to be increasing evidence of cynicism and resistance to choice. The introduction to an article by Meek (2005) about choice in the UK public sector claims "the issue of choice has come to overshadow our everyday lives". As Gawande (1999, p.90) notes "patients commonly prefer to have others make their decisions for them", a phenomenon observed in a number of studies. Magee et al (2003, p.340) report a focus group participant as saying "I shouldn't have to go shopping around to get healthcare for my son. I would hope that my GP would have that sort of information." Further, even where choice "looks attractive in prospect [it] does not always look so good in practice" (Schwartz, 2004, p.104). In a very striking example, Gawande (1999, p.90) reports a study that "found that although sixty-four per cent of the general public thought they'd want to select their own treatment if they developed cancer, only twelve percent of newly diagnosed cancer patients actually did want to do so". Fotaki (1999), in a study of post-operative eye patients in London and Stockholm, found that almost three-quarters were happy for decisions about their treatment to be made by health professionals. Which? (2005) reports a 2004 survey on Quality in Health, which found 69% of respondents agreeing with the statement "I want to be able to choose which healthcare service I use" (p.32), but a 2005 survey on Choices found that 71% of respondents agreed "that patients should trust doctors to make the best decisions for them" (p.36).

The Paradox of Choice

Choosing is difficult – so is choice a good thing? Or is it patronising simply to ask that question? Is choice good for the better-educated and richer sections of society, but not for others? The thought-provoking 2004 book 'The Paradox of Choice' by Barry Schwartz suggests, however, that every section of society might find that choice is not always beneficial.

As noted above, Schwartz (2004, p.99) states unequivocally that "Freedom and autonomy are critical to our well-being, and choice is critical to freedom and autonomy." Nevertheless, elsewhere (p.217) he refers to our "experience of choice as a burden rather than a privilege". He argues (p.104) that "...having the opportunity to choose is no blessing if we feel we do not have the wherewithal to choose wisely..."

It is frequently argued that people will be able to 'choose wisely' if they are provided with relevant information and trained to use it – resulting in 'informed choice'. However, as Appleby et al (2003, p.25) argue, "requirements for informed choice are demanding: not only for the individual, faced with a mass of hard-to-understand information, but also for the healthcare system, which has to provide that information". Even if we can obtain good information, Schwartz (2004, p.56) asks, "do we know how to analyse, sift, weigh, and evaluate it to arrive at the right conclusions and make the right choices?" Further, as we saw in the quotation above from Magee et al (2003), consumers may feel it is not their job to obtain and evaluate the information.

Further, how can consumers judge what is good-quality information? Schwartz (2004, pp 55–6) reported than an assessment by the RAND corporation found that websites providing medical information "with rare exceptions...[were]...all doing an equally poor job", with omissions, and misleading and even inaccurate information. However, even more alarmingly, he reports that "surveys indicate that these web sites actually influence the health-related decisions of 70% of the people who consult them".

Consumers might judge the quality of information by its source, but such judgements can be contradictory. The focus groups reported by Magee et al (2003) were presented with both official Government information, in the form of Department of Health performance ratings, and independent information, in the form of Dr Foster's Good Hospital Guide. However, whilst one respondent is reported as saying "Government figures don't mean anything to me because government is baloney", another stated, "I think that if the Government sent them out then maybe you've got a better chance of believing it" (Magee et al, 2003 p.339). Some found the information in Dr Foster's Good Hospital Guide useful and accessible: for example, "This is much more useful because it is down to basic questions with basic answers, like does the hospital have a newsagent. Little things like that are the sorts of things that it's handy to find out about." However, there was suspicion about the motivation of those supplying such information. One asked "who pays Dr Foster then?" echoed by another who stated, "You don't know who's been the financial backer of something like this. I don't take much information as read. I'm a 'why' girl, I'm afraid." Scepticism also arises when published information does not accord with personal experience, for example, "it is saying that it [Walsgrave Hospital] is a good hospital... and we know it isn't." (p.340)

Further, the way information is interpreted can be unexpected, especially in the light of the Government's emphasis on reducing waiting times. Magee et al (2003, p.340) report a participant of a focus group on choice and comparative performance information as saying: "'If you were told you have got the choice between two and one has a 9-week waiting list and one had a week's waiting list, to be honest I'd go for the 9-week waiting lists because you know that everybody is going there because he's a good guy.'" This is echoed by the findings of Burge et al (2005) who, in a series of choice experiments, found 'reputation' of hospitals to be judged more important than waiting times.

There are other problems relating to the provision of information, which have been widely debated in the literature. For example, the publication of mortality data, even where there is risk-adjustment, could mislead potential patients and also result in hospitals turning away high-risk cases or manipulating the data (Shannon, 2005).

Adverse Effects of Choice

Beyond resistance and difficulties with information, Schwartz (2004, p.74) claims that "the growth of options and opportunities for choice has three, related, unfortunate effects... decisions require more effort, mistakes are more likely and the psychological consequences of mistakes [are] more severe". Drawing on Schwartz, three areas – regret, satisfaction and control – will be explored here.

Regret: Shackley and Ryan (1994, p. 531) define regret as "a psychological relation to making a wrong decision". Schwartz (2004, p.147) argues that whenever "you make a decision and it doesn't turn out well or you find an alternative that would have turned out better, you're a candidate for regret." He stresses that an important determinant of regret is responsibility. Whilst we can be unhappy about a bad decision made by someone else, our regret will be greater if we are responsible for the choice – the responsibility adds to the regret. Shackley and Ryan (1994, p. 531) thus suggest that by "relinquishing the decision-making responsibility to the presumably better informed doctor the possibility of the patient experiencing regret is eliminated". However, despite the fact that they assert that regret "can only be experienced after choice has been made", they suggest that 'anticipated regret' could be important in a patient's decision to cede decision-making to the doctor. Thus we observe regret engendered by choice can arise as the result of a choice being made, but can also occur in advance of any choice being made. Schwartz (2004, p.147) terms these "postdecision regret" and "anticipated regret" and argues (p.148) that both "raise the emotional stakes of decisions".

However, Schwartz also argues that, even if someone makes 'the right' decision, there can be regret because of opportunity costs. He notes (p.133) that "choosing almost always involves giving up something else of value..." and argues that "whenever we are forced to make decisions involving trade-offs, we will feel less good about the option we choose than we would have if the alternative hadn't been there" (p.131).

Further, contrary to arguments that the more choice the better, he states (pp 1.32-3) that "adding options can be detrimental to our well-being ... [because] we don't put rejected options out of our minds...satisfaction with decisions [is] diluted by all the options we considered but did not choose" and thus the "the cumulative opportunity cost of adding options to one's choice set can reduce satisfaction" (p.135).

Further, it is not just the problem of regret itself but, he argues (p.131), the "emotional cost of potential trade-offs.... interferes with the quality of decisions themselves".

Satisfaction/Happiness: Schwartz (2004, p.86) compares maximisers and satisficers, arguing that maximisers are less happy and are more susceptible to all forms of regret than satisficers. Satisficers, he argues, choose something that is 'good enough' and are 'less likely to care if something better is just round the corner'. However, for maximisers, discovering something better "can be a source of real pain". For maximisers, he continued (p.92), overload of choice is nightmare.

Paradoxically, satisfaction does not necessarily appear to relate to the quality of decisions. Schwartz (p.88) concedes that "maximisers might do better objectively than satisficers", but claims that "they tend to do worse subjectively" (p.88). But this leads to the question of what counts – what is important – objective results or subjective experiences?

Rice (2001, p.240) outlines the case for individual autonomy and choices in healthcare. Under the heading of economic efficiency, he suggests that the economic theory of revealed preferences, which "relies on individuals' abilities to make utility-maximising choices", leads to the conclusion that "allowing consumers to make their own choices will lead to the highest level of societal welfare". However, when Rice outlines the case for abridging individual autonomy, he lists instances of market failure in healthcare which could necessitate government involvement and limitation of consumer sovereignty to improve efficiency.

But if, as Schwartz claims, maximisers are less satisfied, why should anyone wish to be a maximiser? Schwartz (2004 p.94) argues firstly that many are not aware that they are maximisers. Secondly, there is concern for status, with "only 'the best'" assuring success when competing against everyone else. This situation can be exacerbated by competition for inherently scarce goods. For example, "the very best medical facilities cannot be made more plentiful" (p.94). But whilst some might inherently be maximisers, Schwartz (p.95) asks whether choice can create maximisers. Is it really possible that greater choice turns otherwise satisfied and content people into dissatisfied maximisers?

Control: Choice is often proposed as enabling people to have greater control over their lives. As Schwartz (p.204) notes, "having control is of crucial importance to psychological well-being" (p204) and "having control over one's life matters" (p206). This is supported by Rice (2001, p.241) who suggests that "individuals are likely to get more satisfaction out of the goods and services they purchase if they choose them" rather than having them assigned to them, even if the goods or services so assigned are those they would have chosen themselves.

However, as suggested above, greater choice – a larger choice set – might not increase control – indeed it might have the opposite effect. As Schwartz (p.104) notes, "more choice may not always mean more control....opportunities become so numerous we feel overwhelmed...". (p.104). Dowding (1992, p.305) suggests not only that the value of adding less-valued alternatives to the choice set is nil, but that, as there are costs associated with making choices, adding 'an option to a choice set might be disadvantageous if it were not the alternative one would choose". This disadvantage would be greater the higher the cost of error – making a wrong decision. "As the number of choices we face increases, freedom of choice eventually becomes a tyranny of choice" (Schwartz 2004 p.235). On the other hand, Dowding (1992, p.306) suggests that the "greater the range of products available the greater the chances that all individuals will have their preferences satisfied".

Control is also related to being given meaningful choice. A 'choice set' that does not contain the chooser's preferred option, or which contains a number of equally bad choices, limits the chooser's control. Possibly more fundamental is control over where choice is permitted – what could be termed the choice 'domain'. There has long been a cynical view of consultations on new road proposals, where the choice offered for consultation is between, say, route A or route B, whilst no debate ('choice') is permitted on whether or not there should be a new road at all. There is considerable evidence from a number of studies (eg Which?, 2005) that the UK public consider availability of good local hospitals and personal, more flexible, services to be more important to them than having a choice of hospital.

Thus, it can be concluded that an important aspect in the control and choice debate is the distinction between being *Forced* to choose, which is disempowering and denies control, and being *Enabled* to choose, which is empowering and enhances control.

What Can be Done?

It is important to stress that the difficulties associated with choice, and the adverse consequences that can arise, cannot be dismissed as applying only to those classified as 'disadvantaged' – who do not have the knowledge or skills to chose – they are inherent features of choice and apply to everyone. There can be few readers of 'The Paradox of Choice' who do not find at least some of the examples echoing their own experience. Whilst the difficulties and adverse effects of choice might affect different people to different degrees, this appears to be related more to their psychology than their educational level or socio-economic status. However, these problems do not mean that choice should be denied or avoided.

Schwartz makes a number of proposals to help cope with the paradox of choice. These include choosing when to choose – we must, he argues (p.222), "decide which choices in our lives really matter". Among other strategies, he recommends (p.225) that we Satisfice more and Maximise less (p.225), and reduce the number of options considered by limiting "the set of possibilities so that the opportunity costs don't add up to make all the alternatives unattractive" (p.133).

Can OR Contribute?

Developing decision aids for patients which encourage Satisficing rather than Maximising could potentially make a major contribution. The approaches of Soft OR (for example Rosenhead & Mingers 2001) would appear to offer considerable scope here, especially when compared to more traditional maximising approaches. However, in moving away from optimising approaches, there needs to be awareness of the potential for regret arising from opportunity costs – from the value of the options forgone.

Is there scope for OR to assist in the constructive limitation of options? The evidence points to the disadvantages of larger choice sets, but at the same time a choice set which excludes attractive options can be considered no choice at all. Can choosers be assisted both in the early elimination of 'no hopers' and the identification of what, to them, are the really important features, so that a 'good enough' option is found, avoiding the regret which might be associated with attempting an evaluation of all options? Paradoxically, however, the choice of any provider in the country, proposed for 2008, might prove less stressful than choosing between four or five options. With the very large choice set, the chooser might be happy, having eliminated one or two hospitals 'they would not touch with a barge pole', to select a hospital on a single criterion, say proximity to family members, without feeling guilt or regret that they did not give proper consideration to all the options.

Another area of potential contribution is assisting decision makers in increasing control over situation. This could include attempting to develop approaches which *enable* choice rather than *force* choice and to enhance control over whether or not to enter a choice situation. These might be possible at an individual level but, more likely, they will need to be tackled at a collective or community level. Some control over the choice set might be possible, but control of the Choice 'domain' – the area where choice is permitted – is another matter. An individual could choose between, say, public hospital A and private hospital B. However, an (ordinary) individual (consumer) cannot choose whether or not the NHS is privatised (or even something rather less contentious such as the location of a community clinic). This inability reflects the limitations of the consumerist 'individual choice' agenda – such decisions are in the political and collective arena. OR, and especially Soft OR, has a lot to offer in promoting and facilitating participation and collective choice, but this is outside the scope of this paper.

Conclusions

Clearly choice, especially in healthcare, is multi-faceted and raises a number of conceptual and practical issues which are generating a vast literature. According to Schwartz (2004) the paradox of choice applies to all – not just some disadvantaged sections of the community. Thus, challenging 'choice', especially where it appears to be being introduced for choice's sake, should not be viewed as patronising.

But, accepting that, what role should OR play in this? Should OR consider choice as a neutral concept and simply offer its technical assistance to the different parties? Or should OR seek to address the negative consequences of choice and attempt to contribute to the exposure, understanding and, if possible, mitigation of them?

References

Appleby J and Dixon J (2004) Patient choice in the NHS: Having choice may not improve health outcomes. *British Medical Journal* 329(7457):61–62

Appleby J, Harrison A and Devlin N (2003) *What is the Real Cost of More Patient Choice?* Kings Fund: London

Burge P, Devlin N, Appleby J, Rohr C and Grant J (2005) *London Patient Choice Project Evaluation.* Rand, King's Fund and City University London

Charnock N (2004) Email consultations in health care: Patients need to be given a choice. *British Medical Journal* 329(7473):1046

Cole A (2005) One in three NHS trusts face cash shortfall this financial year. *British Medical Journal* 331(7507):10

Cook L (2003) In search of a good death: Law needs to be changed to allow terminally ill people choice of a dignified death. *British Medical Journal* 327(7408):225–226

Coombes R (2005) Private providers must be stopped from skimming off easy cases. *British Medical Journal* 330(7493):691

Damiani M, Propper C and Dixon J (2005) Mapping choice in the NHS: cross sectional study of routinely collected data. *British Medical Journal* 330(7486)284–8

Dawson D, Jacobs R, Martin S and Smith P (2004) Is Patient Choice an Effective Mechanism to Reduce Waiting Times? *Applied Health Economics and Health Policy* 3(4):195–203

Deeming C and Keen J (2004) Choice and equity: lessons from long term care. *British Medical Journal* 328(7453):1389–90

DoH (2002) *Reforming NHS Financial Flows - Introducing payment by results.* Department of Health: London.

DoH (2004) *The NHS Improvement Plan* Cm 6268, HMSO:London

DoH (2005a) *Creating a Patient-led NHS – Delivering the NHS Improvement Plan.* Department of Health: London

DoH (2005b) *Introduction to Payment by Results.* Department of Health Website (www.doh.gov.uk accessed 29/7/05)

DoH (2005c) *Commissioning a Patient-Led NHS,* Letter to NHS Chief Executives, Department of Health 28/7/05

Fotaki M (1999) The impact of market oriented reforms on choice and information: a case study of cataract survey in outer London and Stockholm. *Social Science and Medicine* 48(10):1415–1432

Iles A (2003) Giving patients choice puts more pressure on GPs, survey shows. *British Medical Journal* 327(7414):522

Kare-Silver N de (2005) Choose and book – whose choice is it anyway? *British Medical Journal* 330(7499):1093

King's Fund (2003) *Response to: 'Fair and personal to you: Choice, responsiveness and equity in the NHS and social care' A Department of Health consultation document* King's Fund: London

Larcher V (2005) Consent, competence, and confidentiality. *British Medical Journal* 330(7487):353–356

Lewis R (2005) More Patient Choice in England's National Health Service. *International Journal of Health Services* 35(3):479–483

Magee H, Davis L-J and Coulter A (2003) Public views on healthcare performance indicators and patient choice. *Journal of the Royal Society of Medicine* 96(8):338–342

Meek J (2005) It's a tough one... *The Guardian* 2.5.05

Mullen P and Spurgeon P (2000) *Priority Setting and the Public* Radcliffe: Oxon

Rice T (2001) Individual autonomy and state involvement in health care. *Journal of Medical Ethics* 27(4):240–244

Rosen R, Curry N and Florin D (2005) *Public views on choices in health and health care* King's Fund: London

Rosenhead R & Mingers J (eds) (2001) *Rational Analysis in a Problematic World Revisited.* Wiley: Chichester

Schwartz B (2004) *Paradox of Choice.* HarperCollins: New York

Shannon C (2005) London hospital publishes mortality data for individual surgeons. *British Medical Journal* 330(7488):384

Sharpe MC (2002) Discomfort of patient power. *British Medical Journal* 324(7347):1215

Which? (2005) *Which Choice: Health* Policy Report, Which? London

DISEASE MODELLING

CAN PATIENT PROGRESS MODELLING INFORM THE MANAGEMENT OF CANCER PATIENTS?

MARTIN UTLEY[1], STEVE GALLIVAN[1], MARK JIT[1], CHRISTOS PASCHALIDES1, CAROL TAN[2] and TOM TREASURE[2]
[1]*Clinical Operational Research Unit, University College London, UK*
[2]*Department of Cardiothoracic Surgery, Guy's Hospital, London, UK*

Abstract

Patient progress modelling was originally developed in the context of the analysis of leukaemia drug trials. Typically, such modelling involves determining a finite number of different health states in which a patient might be at various times, deriving appropriate sojourn time distributions and transition probabilities to reflect the transfers made between these and then the derivation of time-dependent state probabilities.

We describe the development of a "calculation engine" that can be used within Microsoft Excel to facilitate such modelling and its potential application concerning the use of chemotherapy following curative surgery for lung cancer.

Keywords: mathematical modelling; patient progress modelling; lung cancer.

Introduction

Randomised controlled trials are widely recognised as the "gold standard" method for evaluating new treatments or changes in the management of a group of patients. They involve collecting clinical data in accordance with a strict prescribed protocol. The data concern patients allocated in an unbiased fashion to two or more groups. Pre-specified statistical analysis of these data is designed to answer a single research question; trials with a large number of patients can be designed to address more than one question, but these are atypical.

Typically, a trial will compare a new treatment for a given condition to the existing treatment with respect to outcome. Trial design should be informed by current evidence. That said it is commonly and justifiably argued that, in relying on prospective empirical data, trials are to an extent wasteful of existing knowledge; the statistical analysis involved takes no account of the clinical understanding of the disease processes underlying the chosen outcome measure, nor the wealth of data that might already exist concerning the experience of patients that have received the current treatment.

One instance in the 1970s where this drawback to the randomised controlled trial became apparent was in the context of a disease called Acute Myeloid Leukaemia (AML). At the time, a lot of new chemotherapy drugs and treatment protocols were becoming available and there was a need to evaluate them. However, the time taken to recruit sufficient patients to traditional trials was proving prohibitive. Jackson and Aspden (1979) proposed a different evaluation methodology. They argued that, given the years of experience of treating AML, there was already a good understanding of how patients progressed through the course of their disease. In collaboration with clinical experts, Jackson developed a stochastic model and calibrated it using existing clinical data. The central idea was to carry out trials assigning all consenting patients to the test treatment. Instead of having a group of patients allocated to a control arm, the patient progress model was used to predict how patients on the test treatment would have progressed had they received standard treatment.

Jackson went on to apply patient progress modelling in a number of other clinical contexts (Jackson et al 1981, Aspden et al 1981, Pearce et al 1993). In this paper we discuss patient progress modelling in general terms and briefly describe the development of a computer tool that has been developed to perform the calculations often required for such modelling. An example is then given of a recent application of patient progress modelling that augments information from randomised controlled trials concerning the use of chemotherapy following potentially curative surgery for lung cancer.

An Overview of Patient Progress Modelling

Essentially, patient progress modelling involves deriving time dependent state probabilities that reflect how a patient moves between different health states over time.

The health states defined for the patient progress model of AML developed by Jackson (1979) are shown in figure 1, with patients divided into 2 streams: those dying having not achieved clinical remission and those experiencing a period of remission prior to relapsing and subsequently dying. At the time of Jackson's model, patients with AML did not survive long enough to die from something else.

To generate predictions using such models, mathematical analysis based on probability theory is applied to calculate the probability of an individual patient being in any given health state at any given time. Outcome measures of interest are then estimated based on these time-varying state probabilities.

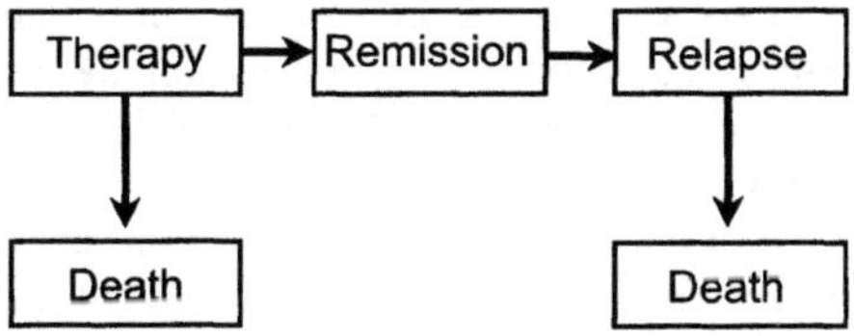

Figure 1: The distinct health states used in the Jackson and Aspden patient progress model of acute myeloid leukaemia.

In the case of the AML study, this approach was used to predict the number of patients in remission over time under the assumption that the trial treatment was no more effective than the existing treatment (see figure 2). The divergence between the trial data and these predictions indicated that the trial treatment was an improvement.

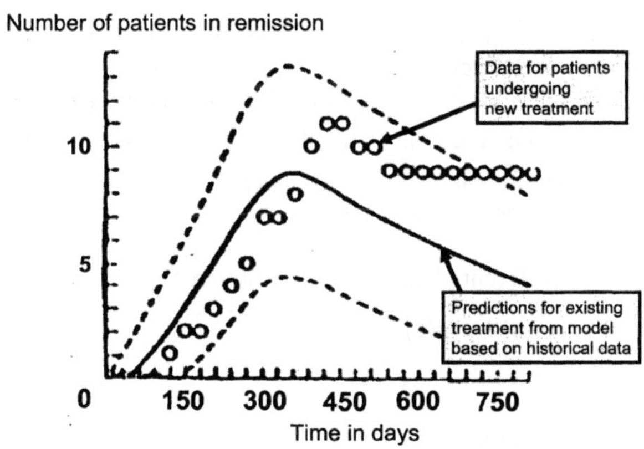

Figure 2: Example results from the Jackson and Aspden evaluation of a chemotherapy treatment for acute myeloid leukaemia.

A Calculation Engine for Patient Progress Modelling

For a model with a relatively simple structure, such as that shown in figure 1, the required analysis can be done by hand. However, in some instances the most suitable structure of health states and transitions proves too complicated for this approach and computer software is needed to carry out the analysis. Contexts in which this is the case include comparing options for the structure of cervical cancer screening programmes (Sherlaw-Johnson et al 1994) and quantifying the health impact of the environmental changes associated with a waste incinerator (McCarthy and Utley 2004). The latter application concerned changes in environmental exposures that affect health in a number of ways. The health effects considered within the patient progress model adopted were chronic bronchitis, cancer, road traffic accidents and other causes of mortality. This gave rise to a model structure with 81 states.

Explicitly or implicitly, such models depend on solutions to linear systems of differential equations linking vectors whose components are probabilities. In vector and matrix notation, these can be expressed in the following form

$$\frac{d\mathbf{p}(t)}{dt} = \mathbf{A}\mathbf{p}(t) \qquad (1)$$

where $\mathbf{p}(t)$ is a vector whose components correspond to the probabilities that a patient is in one of h different health states (perhaps including dummy states) at time t and $\mathbf{A}$ is an h x h matrix with zero column-sums and non-negative entries off the diagonal. Each element a_{ij} $(i \neq j)$ of the matrix $\mathbf{A}$ represents the hazard for moving from health state j to health state i (for an introduction to hazard modelling, see for example Collett 1994).

The solution to the differential equation (1) is given by

$$\mathbf{p}(t+\tau) = \exp(\mathbf{A}\tau)\mathbf{p}(t) \qquad (2)$$

where

$$\exp(\mathbf{A}\tau) = \sum_{n=0}^{\infty} \frac{(\mathbf{A}\tau)^n}{n!}. \qquad (3)$$

For instances where the distribution of sojourn time for a given health state cannot be approximated by an exponential distribution, "dummy" states can be used. The sojourn time distribution of each dummy state is exponential and chosen such that the distribution of the time taken to pass sequentially through all of the dummy states associated with a given health state matches the sojourn time distribution of the physical health state (Harrison and Millard 1991, Harrison 1994, Taylor et al 1998).

Although equation (3) has an elegant and compact mathematical form, it is not straightforward to evaluate this expression directly, indeed it is not possible to evaluate this expression exactly. In order to assist the development of patient progress models we have devised an algorithm for estimating the solution of (3) that can be used as a "calculation engine" in the context of patient progress modelling. Whilst commercially available mathematical analysis packages such as Mathematica (Wolfram Research 2003) incorporate such a feature, reliance on such specialist software packages constrains the type of modelling that can be achieved and the portability of the resulting software. The algorithm that has been developed is suitable for use within the Excel spreadsheet environment (Microsoft 2003). The algorithm used and computer code for applying the algorithm is described in detail elsewhere (Gallivan et al 2006) and an example application is available on the Clinical Operational Research Unit website (http://www.ucl.ac.uk/operational-research/expmat.htm accessed September 2005).

Estimating the Benefit of Post-Operative Chemotherapy in Lung Cancer

Many randomised controlled trials have been conducted to evaluate the use of adjuvant chemotherapy following surgery for non-small cell lung cancer. Meta-analysis of these trials has led to the conclusion that there is a statistically significant survival benefit associated with post-operative chemotherapy (Sedrekyan 2004) and the UK National Institute for Clinical Excellence recommend that the option of post-operative chemotherapy is discussed with the patient (www.nice.org.uk/pdf/cg024fullguideline.pdf accessed 28th September 2005).

The decision concerning whether to follow a major thoracic operation such as a lobectomy with a period of chemotherapy treatment can be a difficult one due to the many side effects associated with chemotherapy (www.cancerhelp.org.uk accessed September 2005).

Whilst giving an estimate of the average benefit to a broad (but not fully representative) population of patients, the trial evidence is of limited use in informing the discussion between an individual patient and a surgeon. In terms of prognosis, the patient population is by no means homogeneous and many patient specific factors are associated with patient survival. Chief amongst these is the stage of the patient's cancer as determined at the time of the operation (Mountain et al 1997). To augment the trial evidence, patient progress modelling has been used to estimate the survival benefit that a patient can expect from post-operative chemotherapy given the stage of the patient's cancer.

Patients face deciding whether to experience several weeks of sustained nausea and fatigue, having just recovered from major surgery and at a time when they are coming to terms with any new information concerning their condition that arose at the time of surgery. It is our opinion that any improvement in the information available to a patient concerning the potential survival benefits of taking this course of action is valuable.

A simplified diagram of the distinct health states considered within the model is shown in figure 3. Within the model a patient either receives or does not receive post-operative chemotherapy and either dies within five years or survives beyond five years. We use the model to estimate outcome measures that relate to the first five years following surgery as this reflects the current focus of lung cancer care in the UK.

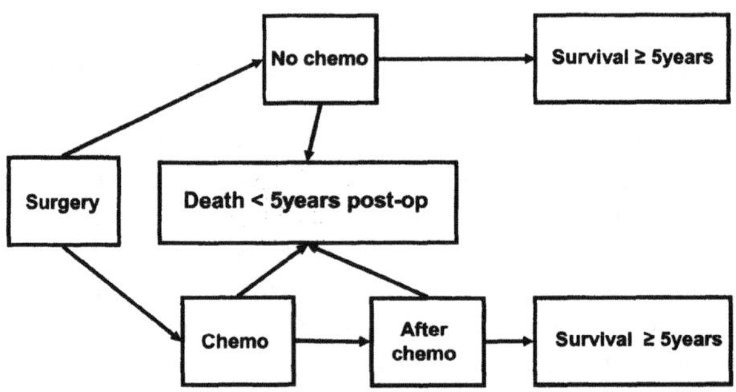

Figure 3: The distinct states used in a patient progress model for estimating the benefits of adjuvant chemotherapy following curative surgery for lung cancer.

Existing data concerning the stage-dependent survival of non-small cell lung cancer patients following surgery was taken from the literature (Mountain et al 1997). To estimate the effect of adjuvant chemotherapy, the value of the hazard ratio to apply in a proportional hazard model (Collett 1994) was taken from a recent meta-analysis (Sedrakyan 2004). The exact form of the patient progress model used in this instance can be found in Utley et al 2006.

The outcome measures of interest are the expected number of weeks of additional survival over five years conferred by post-operative chemotherapy and the percentage chance of surviving beyond five years with and without post-operative chemotherapy. To illustrate the output produced by the model, we use the example of a patient with stage IIIa disease. For such a patient, the estimated chance of surviving beyond five years is 29% if the patient chooses to undergo post-operative chemotherapy, as opposed to 24% if the patient declines chemotherapy. The expected additional survival over the first five years offered by chemotherapy is estimated as 12 weeks. These results should be viewed in the context that post-operative chemotherapy typically involves 3 bouts of week-long therapy over a 9–12 week period.

For many patients recovering from a major operation, these estimated benefits (particularly the expected additional survival over 5 years compared to the duration of chemotherapy) would seem insufficient to outweigh the side affects associated with chemotherapy. That said, other patients may choose to accept a period of significantly worse quality of life in exchange for the expectation of prolonging life and an increase, small as it may seem, in the chance of long term survival. The comparatively small scale of the benefits associated with chemotherapy makes the decision faced by patients all the more difficult and hence any improvements in the information available to patients all the more valuable.

Conclusion

Patient progress modelling has been widely applied to problems in health care. To facilitate the adoption of this modelling approach we have developed and made freely available a "calculation engine" that can be used to solve the matrix differential equations that underpin such modelling.

The attraction of using patient progress modelling to assist in the evaluation of new treatments or changes in the management of patients, is that greater use is made of existing evidence and experience than is the case with randomised controlled trials. This is highlighted by the modelling exercise aimed at evaluating the use of post-operative chemotherapy, in which use is made of existing evidence concerning stage dependent survival. In this instance, the aim has been to augment the information emerging from recent empirical studies. In other contexts, such as assessing options for the management of patients with mesothelioma, patient progress modelling may have the potential to provide useful information in the absence of data from randomised controlled trials. Indeed a modelling study is planned, the aim of which is to provide broad predictive estimates of the results of an ongoing trial (Treasure et al 2004) in this context.

References

Aspden P, Sutcliffe SBJ, Timothy AR and Jackson RRP (1981). An approach to treatment evaluation of Hodgkin's disease. *Omega* 9:537–546.

Collett D (1994). Modelling survival data in medical research. Chapman and Hall: London.

Gallivan S, Utley M, Jit M and Pagel C (2006). A computational algorithm associated with patient progress modelling. *Comp Manag Sci.* in press. http://dx.doi.org/10.1007/s10287-006-0024-x

Harrison GW and Millard PH (1991). Balancing acute and long-stay care: the mathematics of throughput in departments of geriatric medicine. *Methods Inf Med* 30: 221–228.

Harrison GW (1994). Compartmental models of hospital patient occupancy patterns. In: Millard PH and McClean SI (eds). *Modelling Hospital Resource Use.* The Royal Society of Medicine Press, London, pp. 53–61.

Jackson RRP and Aspden P (1979). Treatment evaluation – a modelling approach with application to acute myeloid leukaemia. *J Oper Res Soc* 30:11–22.

Jackson RRP, McKay VL and Malpas JS (1981). Patient progress modelling and results in Wilm's tumour. *Omega* 9:527–536.

McCarthy M and Utley M (2004). Quantitative approaches to HIA. In: Kemm J, Parry J and Palmer S (eds). *Health Impact Assessment.* Oxford University Press, Oxford, pp 61–70.

Microsoft Corporation (2003). *Office Excel 2003 (11.5604.5606).* Redmond, WA.

Mountain CF, Libshitz HI and HE Hermes (1997). "Lung cancer handbook for staging and imaging." Mountain and Libshitz: Houston.

Pearce RM, Gallivan S and Jackson RRP (1993) Patient progress modelling for small cell lung cancer. *Eur J Cancer* 29(Suppl 5): 734–737.

Sedrakyan A, van Der Meulen J, O'Byrne K, Prendiville J, Hill J, Treasure T (2004). Postoperative chemotherapy for non-small cell lung cancer: a systematic review and meta-analysis. *J Thorac Cardiovasc Surg* 128: 414–419.

Sherlaw-Johnson C, Gallivan S, Jenkins D and Jones M (1994). Cytological screening and management of abnormalities in prevention of cervical cancer. *J Clin Pathol* 47: 430–435.

Taylor GJ, McClean SI and Millard PH (1998). Using a continuous-time Markov model with Poisson arrivals to describe the movements of geriatric patients. *Appl Stochastic Models Data Anal* 14:165–174.

Treasure T, Waller D, Swift S, Peto J (2004). Radical surgery for mesothelioma, *BMJ* 328: 237–238.

Utley M, Paschalides C, Treasure T (2006). Informing decisions concerning adjuvant chemotherapy following surgical resection for non-small cell lung cancer: a mathematical modelling study. *Lung Cancer* 53:153–156.

Wolfram Research, Inc. (2003). *Mathematica*, Version 5.0. Champaign, IL.

A SCENARIO-BASED APPROACH TO THE IMPROVEMENT OF HEALTHCARE DELIVERY TO STROKE PATIENTS IN THE COMMUNITY

BESTE KUCUKYAZICI[1], VEDAT VERTER[1],
LYNE NADEAU[2] and NANCY E MAYO[2]
[1]*Desautels Faculty of Management, McGill University, Montreal, Quebec, Canada.*
[2]*Royal Victoria Hospital, Division of Clinical Epidemiology, Montreal, Quebec, Canada.*

Abstract

The objective of this study is to outline a Markov chain model to help to understand, assess and investigate the potentials for improving community-based care delivery processes for stroke patients in the immediate post-discharge period. By using this model, we assess the current system due to the total workload for each health care institution, the expected cost per patient and define the critical paths due to the mortality percentages. We also explore hypothetical scenarios under these quality criteria. Discussion of hypothetical scenarios provides evidence that the model is an effective tool to assess the suggested improvements in the system and investigate potential changes. This approach would therefore seem to be useful in studying the patient paths in the community for any chronic illness.

Keywords: Markov chains; stroke; community-based care delivery process.

Introduction

Patient flow, the progression of patients through health care institutions, is an operational factor in the success of the delivery. Flow paths can provide an insight into both providers and patients about specific health care needs (Cote and Stein, 2000) and support care providers' resource allocation procedures (Thomas, 1968). The application of stochastic models in this domain contributes significantly to an improved understanding and appreciation of patient flow. These models are also beneficial because they can explore the effects of changing decisions about patient care and treatment, on resource use and costs (Davies and Davies, 1994). In this study, the principle outlines of a Markov model are described to understand, assess, as well as investigate, the potentials of integration to improve the community-based care delivery process of stroke patients.

A wide variety of applications of Markov model can be found in health care modeling literature (Akkerman and Knip, 2004; Chaussalet and Thompson, 2001; Christodoulou and Taylor, 2001; Gunes et al., 2004; Sugar, et al. 2004; Zon and Kommer, 1999). It is also one of the popular types of models for patient flows due to its simplicity and efficiency in testing and running the program.

Fetter and Thompson (1968) introduced the modeling of patient flow in a progressive patient-care hospital. Exploiting its simple structure, the use of a Markov chain to approximate patient movement within a hospital has been studied by Thomas (1969). Thomas developed a Markov chain model that was adequate to model certain summary measures of the movement of coronary patients through a care facility. Kao (1974) expanded Thomas' work and used a semi-Markov model to describe the pattern of patients. In previous papers, a state was defined by a specific set of care requirements, such as nursing-skill requirements, available facilities and equipments. In contrast to those, Kao (1974) considered the state definition based solely on the care unit in which the patients reside. Then, several applications on different types of diseases were studied (Cote and Stein, 2000; Davies et al., 1975; Weiss et al., 1982).

In Markov models of patient-paths, the trajectories representing the different sequences of care units traversed by the patients are 'folded' in a compact form through the use of transition-probability matrices. This simplification is beneficial when simulation or analytical techniques are applied to study the related planning and control problems (Kao, 1974). The limiting assumptions are that the future movement of a patient following a given process depends only on the present state (Weiss et al., 1982) and the group of patients is homogeneous (Davies and Davies, 1994).

Until now, all studies that model the patient flows consider the flow through a health care facility; however, they do not account for care delivery paths in the community. To our knowledge, our study is the first attempt that focuses on this area. In this study, an absorbing Markov model of community-based care delivery process for strokes is presented. We use field data to estimate the parameters of the underlying process and evaluate the performance of such a model in estimating the future paths.

Methods

A stochastic process governing the movement of a stroke patient from one care provider to another is considered throughout our study. Whenever the patient visits a care provider i, the next care provider that s/he visits is selected with probability p_{ij}. The transition probabilities p_{ij} form a Markov chain with a transition probability matrix P denoting $\{p_{ij}\}$. It is assumed that the process has a total of q states with t transient states and q-t absorbing states, states where the system reaches one of those states, it never leaves that state (e.g. death).

The q x q transition probability matrix is represented as follows:

$$P = \begin{array}{c:c|l} q & q-t & \\ \left| \; A \right. & \left. B \; \right| & t \\ \hdashline \left| \; 0 \right. & \left. I \; \right| & q-t \end{array}$$

where sub-matrix A describes the movements of patients from transient states to transient states, sub-matrix B describes the movement of patients from transient states to absorbing states, 0 consists of zeros, and sub-matrix I is the identity matrix.

Data

The data set in this study has been obtained from Regie d'assurance maladie du Quebec (RAMQ). All care provider paths following the patients' discharge from Quebec acute-care hospitals in 2001 and their frequencies are determined. While determining the paths, the data obtained from the records of fee-for-service billings on visits to any type of care providers, emergency room and re-admissions after discharge are used. It is decided that 3 months is a reasonable time period for immediate post-discharge in the analysis.

In 2001, a total of 4,317 stroke patients were discharged but 234 of them were back to an acute care hospital within 2 days. These patients are excluded. In the analysis, it is assumed that the care provider path that the patient follows is independent of the demographic variables of the patient such as age, gender or region s/he lives. Because of the lack of comprehensive data, it has not been possible to test this assumption fully. The care providers that the patient visits are grouped in five institutions: family physician, general practitioner, specialist, emergency room, and short-term hospitalization. The way the source data are organized suggests that the path of the patient movement starts when a patient is discharged from hospital. For representing the initialization of the movement, a dummy state, discharge, is defined. All the paths start with dummy state and end with the death of the patient, returning to the long-term hospitalization or visiting any more care provider within 3-months period. The analysis gives that there are 1,279 different types of paths that are followed. Average number of visits to care providers, weighted with the frequency of the paths, is 4.14 and standard deviation is 1.32. We find out the reasonable number of visits as 10 by calculating the value of average number of visits plus 4 times of standard deviation. The paths that include more than 10 visits are evaluated as extreme paths and excluded from the analysis. By this, the total number of effective patients that we evaluate their care patterns reduces to 3,946.

Implementation of Model

The most restrictive assumption of Markov models is that the future movement of a patient depends only on the current state, but independent from the past movements of the patient. However, it is conceivable that this assumption may not hold for patients suffering from certain types of diseases (Weiss et al., 1982). First, the states are defined under the assumption that the care provider the patient visits is independent of previous visits. We define six transient states: discharge from hospital (D), family physician (F), general practitioner (G), specialist (S), emergency room (E), and short-term hospitalization (H). We also define three absorbing states that are death (M), return to long-term hospitalization (LH) and stopping to visit any care provider (END). In the next step, this assumption is tested. If the statement did not hold, a new state space definition that accounts for past history would be necessary.

While testing whether the transitions are Markovian, we apply the procedure proposed by Weiss, Cohen and Hershey (1982). The hypothesis for the Markovian constraint (independency of the transition probabilities) is stated for each state as follows:

Ho: $p_{1ij} = p_{2ij} = \ldots = p_{tij}$ for $i=1,..,t$ and $j=1,..,q$

where

p_{kij}= the probability that the patient reaches j after visiting providers k and i respectively, for $k = 1,\ldots, t$; $i = 1,\ldots, t$; $j = 1,.., q$

The likelihood ratio criterion was used to test these hypotheses and suggested that the null hypothesis should be rejected. This implies that the original Markovian model was inappropriate. We developed a new Markov model by creating new states, which incorporate past history. We redefined the states by adding the memory of the immediate preceding care provider, such as visiting a general practitioner first after discharge from hospital (DG), visiting a general practitioner with the immediate past visit being to a specialist (SG) or no more visits after visiting a general practitioner (G_END). In the new state space, we have a total of 49 states (18 absorbing and 31 transient) describing the movements of the patients. The independency test on the transitions was repeated for the redefined states and suggested that the Markovian property holds for the system with the new set of states, which keep one-period history information.

We modelled 12,717 visits and 15,757 movements (including transitions to the absorbing states) of 3,946 patients in our transition probability matrices. The transition probabilities in the matrix are estimated from the data as follows:

$p_{ij} = n_{ij/} n_i$ for $i = 1,\ldots,m;\ j = 1,.., r$;
n_{ij} = the number of transitions observed going from state i to state j;
n_i = the number of transitions observed in state i.

Scenario-Based Analyses

The proposed patient flow model produces important inputs for planning and control decisions on community-based care delivery process. After validating our Markov model, we conducted scenario analyses so as to develop a solid understanding of the main drivers of health outcomes, workload and costs in the community-based care delivery system to stroke patients.

In analyzing health outcomes, we particularly focus on five sub-paths H-H, G-H, S-H, E-H and H-S, which are associated with significantly higher mortality rates than the overall system average. If the patient were to follow any of these five critical sub-paths the probability of death increases above 10%, while the average is around 2%. Therefore, in our opinion, mitigating the occurrence of these critical sub-paths would result in better health outcomes.

In the scenarios, it is proposed that some improvements could be made in the pathways by re-directing people. It is thought that making a change in the first visits is much more realistic and we define 9 reasonable cases. In each case, we assume that 25% of the patients, who visit care provider X in their first visit, are directed to the care provider Y rather than X. Each of the scenarios was implemented by altering only the associated portion of the transition-probability matrix. Our findings for these scenarios are summarized in Table 1.

The analysis shows us important insights about the dynamics of the patterns. The key findings of the scenario analysis can be summarized as:

1. **Channeling 25% of the patients that go to a specialist during their first visit after discharge, to a family physician (Scenario #1):** This scenario does not reflect a change in workloads for the overall system. This shows us that no matter how many patients, who visit a specialist, are diverted to a family physician, they will all need to visit a specialist, either at the beginning or at other stages of their treatment. In this context, it is obvious that specialists are key elements of the community care delivery process for strokes.

2. **Channeling 25% of the patients that go to specialist during their first visit after discharge, to general practitioner (Scenario #2) and channeling 25% of the patients that go to family physician during their first visit after discharge, to a specialist (Scenario #3):** no significant improvement is observed in the scenarios we studied. The results are as expected.

3. **Channeling 25% of the patients that go to emergency treatment during their first visit after discharge, to a specialist (Scenario #4):** As depicted in Table 1 this results in only a 7% increase on the total workload of specialists, while reducing the total workload on emergency by 20%. Also, because the visits to critical sub-path E-H are reduced by 24%, the number of short-term hospitalizations (H) is 17% less under this scenario. Since H is the most expensive care-provider, this leads to a 14% cost reduction per stroke patient.

4. **Channeling 25% of the patients that go to emergency during their first visit after discharge to a family physician (Scenario #5) or general practitioner (Scenario #6):** Similar results to Scenario #4 are observed in both cases.

5. **Channeling 25% of the patients that go to short-term hospitalization during their first visit after discharge to specialist (Scenario #7):, family physician (Scenario #8): or general practitioner (Scenario #9):** In contrast with scenarios #4, #5 and #6, no significant improvement is observed in the attributes we studied.

Table 1: Number of patients visiting various care providers, critical sub-paths and expected cost per patient

			CASE 1		CASE 2		CASE 3		CASE 4		CASE 5	
			S —> F		S —> G		F —> S		E —> S		E —> F	
	Cost per Visit[1]	Current	Sce.	Change[3]	Sce.	Change[3]	Sce.	Change[3]	Sce.	Change[3]	Sce.	Change[3]
H	$470/day*7d[(2)] = $3290	727	714.00	0.02	749.14	-3.04	713.00	1.93	606.51	16.57	607.18	16.48
F	$40/visit	4450	4911.00	-0.10	4346.42	2.33	3838.00	13.75	4460.16	-0.23	4830.33	-8.55
G	$40/visit	1613	1602.00	0.01	2612.00	-61.93	1638.00	-1.55	1594.44	1.15	1581.86	1.93
S	$55/visit	4586	4238.00	0.08	3842.00	16.22	5042.00	-9.94	4902.17	-6.89	4606.91	-0.46
E	$140/visit	1341	1349.00	-0.01	1408.68	-5.05	1326.00	1.12	1070.72	20.16	1079.60	19.49
HH		17	17.00	0.00	18.00	-5.88	17.00	0.00	15.44	9.20	15.55	8.52
GH		32	32.00	0.00	51.00	-59.38	33.00	-3.13	30.54	4.56	30.33	5.22
SH		75	69.00	8.00	66.00	12.00	82.00	-9.33	76.62	-2.15	71.69	4.42
EH		497	485.00	2.41	508.00	-2.21	478.00	3.82	376.95	24.16	379.66	23.61
HS		145	1115.00	-9.64	145.00	0.00	143.00	1.38	123.34	14.94	122.45	15.55
Av. Cost		779.1	768.25	1.39	798.66	-2.51	767.30	1.51	673.37	13.57	673.75	13.52

			Case 6		Case 7		Case 8		Case 9		Case 10	
			E —> G		H —> S		H —> F		H —> G		E —> S[4]	
	Cost per Visit[1]	Current	Sce.	Change[3]	Sce.	Change[3]	Sce.	Change[3]	Sce.	Change[3]	Sce.	Change[3]
H	$470/day*x7d[2] = $3290	727	621.21	14.55	679.42	6.55	669.38	7.93	680.78	6.36	568.13	21.85
F	$40/visit	4450	4413.33	0.82	4474.96	-0.56	4505.52	-1.25	4462.00	-0.27	4450.12	0.00
G	$40/visit	1613	2004.36	-24.26	1632.05	-1.18	1615.35	-0.15	1675.99	-3.91	1846.03	-14.45
S	$55/visit	4586	4594.76	-0.19	4671.14	-1.86	4602.93	-0.37	4606.71	-0.45	4881.19	-6.44
E	$140/visit	1341	1097.84	18.13	1343.00	-0.15	1340.85	0.01	1344.69	-0.28	920.85	31.33
HH		17	15.77	7.24	14.01	17.62	13.31	21.71	14.08	17.18	14.99	11.82
GH		32	38.04	-18.86	32.47	-1.48	32.37	-1.16	33.46	-4.56	34.00	-6.25
SH		75	72.97	2.71	75.39	-0.52	74.12	1.17	74.57	0.57	79.00	-5.33
EH		497	388.10	21.91	482.34	2.95	481.44	3.13	483.00	2.82	334.00	32.80
HS		145	124.62	14.06	134.17	7.47	131.75	9.14	134.23	7.43	145.62	-0.43
Av. Cost		779.1	685.98	11.95	741.13	4.87	731.87	6.06	741.74	4.80	638.21	18.08

(1) All costs are in terms of Canadian Dollars (Teng et al., 2003).
(2) It is assumed that average length of stay in short term hospitalization is 7 days.
(3) Negative change indicates an increasing, positive change indicates a decreasing.
(4) Guiding the %25 of the patients who visit emergency in their second or further visits to the specialist

6. **Channeling 25% of the patients that go to emergency <u>after</u> their first visit following discharge to a specialist (Scenario #10):** This scenario is studied in order to investigate whether the first visit show different characteristics than the following visits. As depicted in Table 1 under this scenario the workload of emergency is decreased by 31%, the workload of H is decreased by 22% and average cost is also improved by 18%. Perhaps equally important are the 12% and 33% reduction in the number of visits to critical sub-paths H-H and E-H, respectively.

In the next step, we would like to investigate the potentials to improve the process. In this context, we developed a mathematical model to establish the potentials that we could utilize if we oriented the patients for their first care provider visits. In this model, we define the selection of the first care provider to visit as a variable decision but keep the remaining parts of the transition matrix unchanged. Under various objective functions, such as minimization of total workload on specialists, we ran the program. The results are given in Table 2.

		Objective Function										
	Cur.	Min. G[1]	Min. F[1]	Min. S[1]	Min. H[1]	Min. E[1]	Min. HH[2]	Min. EH[2]	Min. GH[2]	Min. SH[2]	Min. HS[2]	Min. Cost
First Visit		*F*	*G*	*H*	*S*	*H*	*S*	*S*	*F*	*F*	*F*	*F*
G	1613	656	6968	701	854	701	854	854	656	656	656	656
F	4450	7900	1673	1874	2394	1874	2394	2394	7900	7900	7900	7900
S	4586	3189	3006	2474	7671	2474	7671	7671	3189	3189	3189	3189
H	727	405	613	4217	400	618	400	400	405	405	405	405
E	1341	830	1094	618	702	702	702	702	830	830	830	830
HH	17	8	12	325	6	325	6	6	8	8	8	8
EH	497	272	399	258	237	258	237	237	272	272	272	272
GH	32	13	126	19	14	19	14	14	13	13	13	13
SH	75	40	59	120	114	120	114	114	40	40	40	40
HS	145	78	110	955	92	955	92	92	78	78	78	78
Ave. Cost	779.1	498	679	3599	498	3599	498	498	498	498	498	498

[1] minimization of the total workload on related institution
[2] minimization of the number of visits to related sub-path

Table 2: Values of quality criteria under different objective functions

The outputs give an idea of the drawbacks of the current system and the potential for improvement. Within the objective function, in order to minimize the workload of specialists, the model diverts all patients to short-term hospitalization and vice versa. That points outs to us the relationship between these two institutions. It may be concluded that there is a bottleneck formation due to the lack of specialists in the system, which causes those patients, who could not visit specialists, to be headed directly into short term hospitalization.

Moreover, the relationship between the emergency room and short-term hospitalization discussed in the scenario analysis is again emphasized under the objective function of minimizing the workload of the emergency room. In analyzing the outputs for important objective functions such as minimizing short-term hospitalization cases, visiting critical sub-paths or minimizing the average cost, the system is continuously optimized by allocating first visits to a specialist or family physician. This issue can be evaluated as evidence, that the a possible improvement could be established by urging patients to visit a specialist or family physician first.

The scenario analysis and the feedbacks obtained from the optimization model point out to us that the three crucial institutes must be used to maximum efficiency: specialists, family physicians and emergency rooms, in order to improve the system. An increase in the number of specialists and family physicians, and guidance to patients to visit them initially, will help to decrease the number of patients visiting emergency rooms. This will also result in improving the short-term hospitalization process. Furthermore, the overall system will benefit from these adjustments in re respect of all the quality criteria, mentioned in this study.

Conclusions

For achieving an efficient overall system performance, the pattern of care in community following a stroke has a significant impact. For the successful management of these care delivery processes, the understanding and appreciation of patient flows is a critical milestone, as it will enable planning and control decisions can be made more accurately.

In this study, we provide a proof of concept that a Markov chain model is an effective tool for modeling patient paths in the home-care delivery process. Field data is used for the verification and validation of the model. The dependence of flows upon previous care providers visited and definition of new states with which to deal with this dependence, allow the model to provide a richer understanding of the pattern. With this model, based on the last two care providers visited- the probability of a next visit to each of the care providers can be predicted quite accurately.

Discussions on hypothetical scenarios provide evidence that the model is an effective tool to assess the suggested improvements in the system and investigate potentials. This approach would therefore seem to be useful in studying the patient paths in the community for any chronic illness.

Throughout the study, some assumptions are made because of data limitations, such that the transition probabilities are drawn from the same distributions for the subpopulations due to the demographic variables and severity of stroke. However, in the validation, we reach really satisfactory test statistics that help us to figure out that these assumptions are not strong assumptions, which could change the outputs of the model.

In this paper, we describe the principle outlines of the model. As a future work, we are planning to cover the time lags between the visits and to extend the model as a semi-Markov model. It would give a better understanding of the pattern and provide opportunities for a richer analysis.

References

Akkerman R and Knip M (2004). Reallocation of beds to reduce waiting time for cardiac surgery. *Health Care Management Science* 7; 2: 119–130.

Chaussalet J and Thompson A (2001). Data requirement in a model of the nature history of Alzheimer's disease. *Health Care Management Science* 4:13–19.

Christodoulou G and Taylor G J (2001). Using continuous time hidden Markov process, with covariates, to model bed occupancy of people aged over 65 years. *Health Care Management Science* 4:21–24.

Cote M and Stein W E (2000). An Erlang-based stochastic model for patient flow. *Omega* 28:347–359.

Davies and Davies H T O (1994). Modeling patient flows and resource provision in health systems. *Omega* 22; 2: 123–131.

Davies R, Johnson D and Farrow S (1975). Planning patient care with a Markov model. *Operations Research Q* 26: 599–607.

Gunes E D, Chick S E and Aksin O Z (2004). Breast cancer screening services: trade-offs in quality, capacity, outreach and centralization. *Health Care Management Science* 7: 291–303.

Fetter R B and Thompson J D (1968). A decision model for the design and operation of a progressive patient care hospital. *Medical Care* 7: 450–462.

Kao E P C (1974). Modeling the movement of coronary patients within a hospital by semi-Markov processes. *Operations Research* June–August: 683–69.

Sugar C A, James G M, Lenert L A, et al. (2004). Discrete state analysis for interpretation of data from clinical trials. *Medical Care* 42; 2: 183–196.

Teng J, Mayo E N, Latimer E, et al. (2003). Costs and caregiver consequences of early supported discharge for stroke patient. *Stroke* 34: 528–536.

Thomas, W A (1969). A model for predicting recovery process of coronary patients. *Health Services Research* 3: 185–213.

Weiss E N, Cohen M A and Hershey J C (1982). An iterative estimation and validation procedure for specification of semi-Markov models. *Operations Research* 30; 6: 1082–1104.

Zon A H and Kommer G J (1999). Patient flows and optimal health-care resources allocation at the macro-level: a dynamic linear programming approach. *Health Care Management Science* 2; 2: 87–96.

THE DEVELOPMENT OF A RISK MODEL TO ASSIST WITH THE MONITORING OF SURGICAL WOUND INFECTIONS

CHRIS SHERLAW-JOHNSON[1], STEVE GALLIVAN[1] and PETER WILSON[2]
[1]*Clinical Operational Research Unit, University College London, UK*
[2]*Department of Clinical Microbiology, University College London Hospitals, UK*

Abstract

Infections acquired during a hospital stay are a major health care concern for patients in the UK. They can be fatal, lead to excess morbidity and lengthen hospital stay. There is therefore considerable interest in using analytical tools for monitoring the occurrence of infections, so that any problems with the quality of patient care can be quickly identified and rectified. An important feature of such monitoring tools is a mechanism that accounts for variations in patient case-mix. This paper describes the development and validation of a model for assessing the risk of infection in patients about to receive coronary artery bypass grafting.

Keywords: hospital-acquired infections; surgical wounds; performance monitoring; risk models.

Introduction

Routine wound surveillance with feedback is an important method for controlling the rate of surgical wound infections within the UK National Health Service (NHS). Infections increase the risk of death, worsen morbidity and place a burden on hospital resources. Surgical wound infections affect between 6% and 15% of surgical patients and can more than double the cost of care (Coello et al., 2005). Wound surveillance can be assisted by computer tools that enable healthcare professionals to continuously monitor infection rates (Morton et al., 2001). Any notable variations can be quickly identified, allowing timely remedial action to be taken. Such monitoring can also assist in assessing the overall effects of the introduction of new infection control policies.

Risk models are regularly used to assist with the monitoring of healthcare outcomes. Their purpose is to assess a patient's risk of an adverse event prior to the care they receive so that, for example, normal rates of adverse outcomes among a cluster of higher risk patients are not mistaken for anything unusual. Examples of risk models that have been used in this way include the Parsonnet score and EuroSCORE for predicting mortality after coronary artery bypass surgery (CABG) (Parsonnet et al., 1989; Nashef et al., 1999) and a model for predicting mortality after hospital admission for acute myocardial infarction (Dorsch et al. 2001). In this paper we describe the development and validation of a risk model for predicting the occurrence of surgical wound infections among patients about to receive CABG surgery.

Our risk model uses only information that is available before the operation. In this way it does not correct for the quality of surgery or post-operative care, both of which would be under scrutiny in the monitoring process. Previous studies have established positive links between the risk of infection and pre-operative factors such as patient obesity, diabetes and age (Harrington et al., 2004; Kim et al., 2003; Kubal et al., 2005). Also, unsurprisingly, the more wounds required by an operation, the greater the likelihood that one will become infected.

The ability of hospital management to be able to respond quickly to divergent infection rates is hampered by the fact that over 50% of infections may not become apparent until after hospital discharge (Bruce et al., 2001). Furthermore, post-discharge surveillance is not carried out by many hospitals in the UK due to cost. Even if systems of post-discharge surveillance were in place, there may be substantial delays before receiving information. Therefore we have developed a monitoring tool that only assesses infections that are manifest before patients are discharged.

Data

Since April 2000, University College London Hospitals NHS Trust (UCLH) has been collecting surgical wound infection data as part of its infection surveillance programme. The study uses data for 1,652 patients who had undergone a CABG operation between April 2000 and March 2004. The status of each wound was assessed by the ASEPSIS score (Wilson et al., 1986). A wound was classed as infected if the score was greater than 10, and pre-discharge infections could be identified if the score exceeded 10 during the post-operative stay. Using this definition, 270 (16.3%) patients were deemed to have an infection prior to hospital discharge. It should be noted that this will be greater than the true infection rate because an ASEPSIS score higher than 10 is only an indication of a possible infection: only a proportion of wounds with scores between 10 and 20 will actually be infected.

Developing the Model

70% of the 1,652 patient records were selected randomly for developing the risk model. The resulting model was then validated using data for the remaining 30% of patients. Data items were selected from the UCLH database for possible inclusion in the risk model if they reflected information that would be known before the operation took place. Overall pre-operative health was summarised using the American Society of Anaesthesiologists (ASA) Score (Anon, 1963). Pre-operative management was reflected by the number of days between admission and operation. Factors representing the type of operation the patient would undergo were the number of surgical wounds and whether the operation included valve replacement alongside the bypass surgery. Other patient characteristics considered included the body mass index (BMI – defined as patient weight in kilograms divided by the square of their height in metres), age and gender.

The risk model was developed using logistic regression with a forward stepping procedure for selecting and testing the different factors for inclusion in the model (Hosmer and Lemeshow, 2000). Logistic regression is a useful tool for relating multiple factors to binary outcomes, and with forward stepping we can test the contribution that each factor makes to the model by adding one at a time and calculating the log likelihood ratio. If additional factors have no significant contribution to make then they can be excluded from the model. This was done using SPSS version 12.0.1 (SPSS Inc., 2003). Factors were only included if they made a significant increase to the log likelihood function at the 5% level. The resulting model is given by:

$$\text{risk of infection} = \frac{1}{1+e^{z}}$$

where $z = 3.327$

- $-\ 0.425$ if patient is obese (BMI > 30 kg/m^2)
- $-\ 1.158$ if the operation included valve replacement
- $-\ 0.572 \times$ number of wounds
- $-\ 0.422$ if patient aged > 70
- $-\ 0.529$ if there were more than 2 days between admission and surgery.

Predicted risks varied between 6% for patients with one wound and none of the other factors present, to 82% for patients with four wounds and who are positive for all the other factors. Neither gender, ASA score nor the use of an implant had any significant contribution to make to the risk model.

Validating the Model

The risk model was independently validated using data for the 30 % of patients that had not been used in its development. There were 495 patients in this set, but the model could be tested against only 447 of these, since data for certain risk factors were missing (mainly BMI). Use of the model predicted 74 infections among these patients, compared with an actual number of 73. For a more detailed validation, two methods were used. In the first method we stratified the patients into five sets, determined according to increasing risk, and compared the predicted and observed outcomes in each set. The numbers in each set were 157, 58, 79, 78 and 75. It was not possible to achieve strata with broadly equal numbers in each set, since with a relatively small number of risk factors, tied risk scores were frequent. In particular, 111 patients shared the same low risk score. Expected and actual infection rates within the five risk strata are shown in Figure 1. A chi-squared test indicates a good fit between observed and expected results (chi-square = 1.92, p = 0.86).

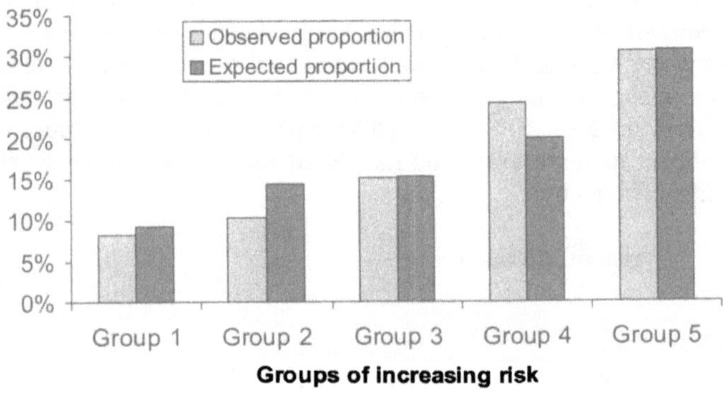

Figure 1: Comparison of observed and expected outcomes among five groups ordered by increasing risk. The number of patients in each group are, respectively, 157, 58, 79, 78 and 75. The associated patient risks are < 12%, 12% to 14.7%, 14.7% to 16.5%, 16.5% to 23%, > 23%.

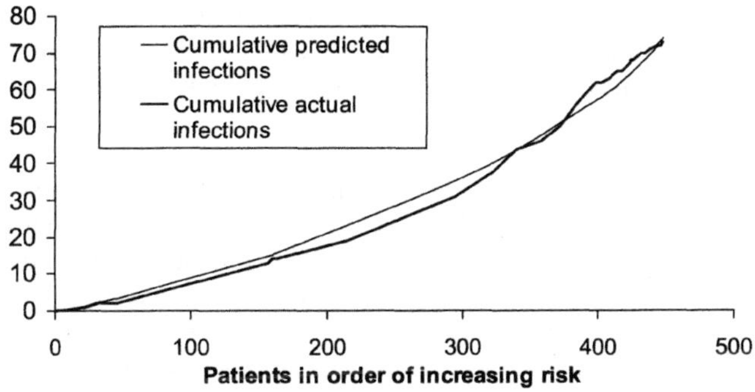

Figure 2: Cumulative observed and expected outcomes over the series of patients ordered by increasing order of risk.

In the other validation method we ordered patients by increasing risk and then plotted cumulative observed and predicted outcomes over this series. Where more than one patient has the same risk score, we extrapolated cumulative observed outcomes linearly over the group. This is shown in figure 2; clearly the two curves are very close indicating that the model makes a good assessment of a patient's risk.

Conclusions

In this study we have developed a risk model for predicting the occurrence of surgical wound infections after CABG operations that become apparent before hospital discharge. The model only includes pre-operative information that is routinely available to the infection control unit at UCLH. For this reason, although other studies have indicated a relationship between diabetes and infection, (Harrington et al., 2004; Kubal et al., 2005), we have not included it.

This study differs from previous analyses of infection following coronary artery bypass surgery in that it sets out to develop a risk model that can be used in routine practice, as opposed to investigating the relationships between individual factors and outcome. Moreover, we have been careful to restrict outcomes to only those infections that are detected prior to hospital discharge. Harrington et al. (2004) also only analysed pre-operative risk factors and found increased age, obesity and diabetes to be independent predictors. Kim et al. (2004) were exclusively interested in the influence of obesity on a variety of outcomes and found it to be a positive predictor of deep sternal wound infection.

On the other hand, Kubal et al. (2005) were exclusively interested in the impact of diabetes, and found a positive relationship with deep sternal wound infection. None of these studies investigated the impact of valve surgery or length of hospital stay prior to surgery.

The model will assist the continuous monitoring of infection rates which, in turn, will enable healthcare professionals to respond quickly to unusual variations in outcome. Risk models can also assist wound surveillance programmes in other ways. At UCLH post-discharge data are also routinely collected, and it would be possible to use them to develop a model for predicting all wound infections that could then be used for retrospective audits. If there is an interest in focussing upon the post-surgical care, relevant surgical information could be tested for possible inclusion in a risk model: for example, the duration of surgery, category of wound contamination and patient status on leaving the operating theatre. Also, since most hospitals do not have the capability to undertake post-discharge surveillance, a model for predicting post-discharge infections would be very useful for identifying those most likely to be needed to be called back to clinic.

References

Anon. New classification of physical status (1963). *Anaesthesiology* 24: 111

Bruce J, Russell EM, Mollison J and Krukowski ZH (2001). The measurement and monitoring of surgical adverse events. *Health Technol Assess* 5 (22): 1–194.

Coello R, Charlett A, Wilson J, Ward V, Pearson A and Borriello P (2005). Adverse impact of surgical site infections in English hospitals. *J Hosp Infect* 60: 93–103.

Dorsch MF, Lawrance RA, Sapsford RJ, Oldham J, Greenwood DC, Jackson BM, Morrell C, Ball SG, Robinson MB, Hall AS and the EMMACE Study Group (2000). A simple benchmark for evaluating quality of care of patients following acute myocardial infarction. *Heart* 86: 150–154.

Harrington G, Russo P, Spelman D, Borrell S, Watson K, Barr W, Martin R, Edmonds D, Cocks J, Greenbough J, Lowe J, Randle L, Castell J, Browne E, Bellis K and Aberline M (2004). Surgical-site infection rates and risk factor analysis in coronary artery bypass graft surgery. *Infect Control Hosp Epidemiol* 25: 472–476.

Hosmer DW and Lemeshow S (2000). *Applied logistic regression. 2nd Edition.* Chichester, New York, Wiley.

Kim J, Hamar N, Jakobsson K, Luepker RV, McGovern PG and Ivert T (2003). Obesity and the risk of early and late mortality after coronary artery bypass graft surgery. *Am Heart J* 146: 555–560.

Kubal C, Srinivasan AK, Grayson AD, Fabri BM and Chalmers JA (2005). Effect of risk-adjusted diabetes on mortality and morbidity after coronary artery bypass surgery. *Ann Thorac Surg* 79: 1570–1576.

Morton AP, Whitby M, McLaws M, Dobson A, McElwain S, Looke D, Stackelroth J and Sartor A (2001). The application of statistical process control charts to the detection and monitoring of hospital-acquired infections. *J Qual Clin Practice* 21: 112–117.

Nashef SAM, Roques F, Gauducheau E, Lemeshow S, Salamon R and the *Euro*SCORE Study Group (1999). European system for cardiac operative risk evaluation (*Euro*SCORE). *Eur J Cardiothorac Surg* 16: 9–13.

Parsonnet V, Dean D and Bernstein AD (1989). A method of uniform stratification of risk for evaluating the results of surgery in acquired adult heart disease. *Circulation* 79 (suppl. I): I3–I12.

SPSS Inc. (2003). *SPSS version 12.0.1 for Windows*. Chicago, Illinois.

Wilson AP, Treasure T, Sturridge MF and Gruneberg RN (1986). A scoring method (ASEPSIS) for postoperative wound infections for use in clinical trials of antibiotic prophylaxis. *Lancet* 1: 311–313.

CORRESPONDING AUTHORS' ADDRESSES

Tobias Andersson
Linköping University
ITN, SE-60174 Norrköping
Sweden.
E-mail: toban@itn.liu.se

Anthony Codrington-Virtue
RKW Healthcare Strategists
146-148 Clerkenwell Road
London EC1R 5DG
UK
E-mail:
Anthony.Codrington-Virtue@rkw-uk.com

Vanda De Angelis
Dipartimento di Statistica e Statistiche Applicate
Università "La Sapienza"
00185 Roma
Italy
E-mail: Vanda.Deangelis@uniroma1.it

Mario Jorge F. De Oliveira
Universidade Federal do Rio de Janeiro
Preograma de Engenharia de Producio
Centro de Tecnologia
Bloco F sala 105
Caixa Postal 68501
Rio de Janeiro, Brazil
E-mail: Mario_jo@pep.ufrj.br

Martin Dlouhy
Faculty of Informatics and Statistics
University of Economics
W. Churchill Sq. 4, 13067 Prague 3
Czech Republic
E-mail: dlouhy@vse.cz

Burcu Toker Ersöz
Industrial Engineering Department
Girne American University
TRNC Mersin 10
Turkey
E-mail: bersoz@gau.edu.tr

Steve Gallivan
Clinical Operational Research Unit
University College London
Gower Street, London, WC1E 6BT
UK
E-mail: s.gallivan@ucl.ac.uk

Murat M. Gunal
Department of Management Science
The Management School
Lancaster University, Lancaster, LA1 4YX
UK.
E-mail: m.gunal@lancaster.ac.uk

Saeedeh Ketabi
Department of Management
University of Isfahan
Iran
E-mail: sketabi@cc.iut.ac.ir

Beste Kucukyazici
Faculty of Management
Mcgill University
1001 Sherbrooke Street West
Montreal, Quebec
Canada H3A 1G5
E-mail: beste.kucukyazici@mail.mcgill.ca

Penelope M Mullen
Health Services Management Centre
40 Edgbaston Park Road
Birmingham
B15 2RT
E-mail: p.m.mullen@bham.ac.uk

Marie Persson
Blekinge Institute of Technology
Department of Software and Systems Engineering
School of Engineering
Box 520
372 25 Ronneby
Sweden
E-mail: marie.persson@bth.se

Barry Shaw
Queen's University of Belfast
Department of Applied Mathematics
David Bates Building
Belfast
Northern Ireland, UK
BT7 1NN
E-mail: barry.shaw@qub.ac.uk

Chris Sherlaw-Johnson
Clinical Operational Research Unit
University College London
Gower Street, London, WC1E 6BT
UK
E-mail: c.sherlaw-johnson@ucl.ac.uk

Martin Utley
Clinical Operational Research Unit
University College London
Gower Street, London, WC1E 6BT
UK
E-mail: m.utley@ucl.ac.uk

Colin Thunhurst
Department of Nursing Midwifery and Healthcare
Faculty of Health and Life Sciences
Coventry University
Coventry CV1 5FB
UK
E-mail: c.thunhurst@coventry.ac.uk

Mario Vanhoucke
Faculty of Economics and Business Administration
Ghent University
Hoveniersberg 24
9000 Gent
Belgium
E-mail: mario.vanhoucke@ugent.be

Dave Worthington
Department of Management Science
Lancaster University Management School
Lancaster, LA1 4YX
UK
E-mail: d.worthington@lancaster.ac.uk

LIST OF PARTICIPANTS

Surname	First Name	Country	E-mail
Abbi	Revlin	United Kingdom	R.Abbi02@westminster.ac.uk
Amado	Carla	Portugal	camado@ualg.pt
Andersson	Tobias	Sweden	toban@itn.liu.se
Baber	Jack	United Kingdom	BaberJMA@cf.ac.uk
Bagust	Adrian	United Kingdom	A.Bagust@liv.ac.uk
Bensley	David	United Kingdom	David.bensley@dh.gsi.gov.uk
Blake	John	Canada	john.blake@dal.ca
Bowen	Tom	United Kingdom	tom.bowen@balanceofcare.com
Brailsford	Sally	United Kingdom	scb@soton.ac.uk
Bruni	Maria	Italy	mebruni@deis.unical.it
Buhaug	Harald	Norway	Harald.Buhaug@helse-midt.no
Carter	Michael	Canada	carter@mie.utoronto.ca
Ceglowski	Andrzej	Australia	red.ceglowski@infotech.monash.edu
Chaussalet	Thierry	United Kingdom	chausst@westminster.ac.uk
Codrington-Virtue	Anthony	United Kingdom	Anthony.Codrington-Virtue@rkw-uk.com
Davies	Ruth	United Kingdom	Ruth.Davies@wbs.ac.uk
De Angelis	Vanda	Italy	Vanda.Deangelis@uniroma1.it
De Oliveira	Mario Jorge	Brazil	Mario_jo@pep.ufrj.br
De Oliveira	Monica	Portugal	Monica.oliveira@tagus.ist.utl.pt
De Souza	Luciano	United Kingdom	l.brandao@lancaster.ac.uk
Delesie	Luc	Belgium	luc.delesie@med.kuleuven.ac.be
Dodds	Simon	United Kingdom	Simon.dodds@goodhope.nhs.uk
Duarte	Alexandre	Brazil	aduarte@inf.puc-rio.br
Ersoz	Burco	Turkey	bersoz@gau.edu.tr
Figlova	Zuzana	Czech Republic	Figlova@vse.cz
Gallivan	Steve	United Kingdom	s.gallivan@ucl.ac.uk
Gunal	Murat	United Kingdom	m.gunal@lancaster.ac.uk
Griffiths	Jeff	United Kingdom	Griffiths@cf.ac.uk
Hans	Erwin	Netherlands	e.w.hans@sms.utwente.nl
Harper	Paul	United Kingdom	prh@soton.ac.uk
Haspeslagh	Marc	Belgium	marc.haspeslagh@med.kuleuven.be
Jaspers	Thieu	Netherlands	Thieu.jaspers@xs4all.nl
Kalai	Rim	France	kalai@lamsade.dauphine.fr
Katsaliaki	Korina	United Kingdom	korina@soton.ac.uk
Ketabi	Saeedeh	Iran	sketabi@cc.iut.ac.ir
Kopach	Renata	USA	rkopach@purdue.edu
Kommer	Geert Jan	Netherlands	G.Kommer@RIVM.nl
Kraus	Marcus	Austria	markus-kraus@aon.at
Kucukyazici	Beste	Canada	bkucuk@po-box.mcgill.ca
Kunene	Niki	USA	knkunene@vcu.edu

Lagergren	Mårten	Sweden	marten.lagergren@aldrecentrum.se
Lamiri	Mehdi	France	Lamiri@emse.fr
Lawson	Zoe	United Kingdom	LawsonZF@cf.ac.uk
Leite Jr	Evandro	United Kingdom	jr@evandro.org
Lins	Marcos	Brazil	lins@pep.ufrj.br
Litvak	Nelly	Netherlands	N.Litvak@ewi.utwente.nl
Lubicz	Marek	Poland	Marek.Lubicz@pwr.wroc.pl
Macfarlane	John	United Kingdom	Johelmacf@aol.com
Mikitis	Eriks	Latvia	eriks@valmiera.lanet.lv
Mullen	Penelope	United Kingdom	p.m.mullen@bham.ac.uk
Ozcan	Yasar	USA	ozcan@vcu.edu
Patel	Brijesh	United Kingdom	B.Patel41@wmin.ac.uk
Patel	Suchi	United Kingdom	Patel03@phd.wbs.ac.uk
Pham	Nguyen	Switzerland	Phamdinh.nguyen@unifr.ch
Person	Marie	Sweden	marie.persson@bth.se
Pitt	Martin	United Kingdom	m.pitt@exeter.ac.uk
Potts	Chris	United Kingdom	C.N.Potts@soton.ac.uk
Powell	Naomi	United Kingdom	N.H.Powell@soton.ac.uk
Price-Lloyd	Naomi	United Kingdom	PricelloydN@cf.ac.uk
Rauner	Marion	Austria	marion.rauner@univie.ac.at
Royston	Geoff	United Kingdom	geoff.royston@dh.gsi.gov.uk
Salustri	Linda	Italy	lindasalustri@hotmail.com
Sbandi	Paola	Italy	paoletta379@yahoo.it
Shahani	Arjan	United Kingdom	akshahani@hotmail.com
Shaw	Barry	United Kingdom	barry.shaw@qub.ac.uk
Sherlaw-Johnson	Chris	United Kingdom	c.sherlaw-johnson@ucl.ac.uk
Smith	Honora	United Kingdom	hks102@soton.ac.uk
Sykes	Jenni	United Kingdom	J.Sykes@soton.ac.uk
Thunhurst	Colin	Ireland	c.thunhurst@ucc.ie
Utley	Martin	United Kingdom	m.utley@ucl.ac.uk
Van Berkel	Peter	Canada	Peter.vanberkel@dal.ca
Vanhoucke	Mario	Belgium	Mario.vanhoucke@UGent.be
Vasilakis	Christos	Canada	christos@interchange.ubc.ca
Vissers	Jan	Netherlands	Vissers@bmg.eur.nl
Williams	Janet	United Kingdom	WilliamsJE@cf.ac.uk
Worthington	Dave	United Kingdom	d.worthington@lancaster.ac.uk
Wullink	Gerhard	Netherlands	g.wullink@erasmusmc.nl
Xie	Xiaolan	France	xie@emse.fr
Yevseyeva	Iryna	Finland	iyevsev@cc.jyu.fi
Zeppelzauer	Wolfgang	Austria	wolfgang.zeppelzauer@univie.ac.at

PUBLISHED PROCEEDINGS FROM PREVIOUS MEETINGS

Matson, E. (Ed), Operational Research Applied to Health Services, Proceedings of the 23rd Annual Meeting of the EURO Working Group on Operational Research Applied to Health Services 1997, Norwegian University of Science and Technology, Trondheim, 1997.

De Angelis, V., Ricciardi, N. and Storchi, G. (Eds), Monitoring, Evaluating, Planning Health Services, Proceedings of the 24th Annual Meeting of the EURO Working Group on Operational Research Applied to Health Services 1998, World Scientific, Singapore, 1998.

Mikitis, E. (Ed), Information, Management and Planning of Health Services, Proceedings of the 25th Annual Meeting of the EURO Working Group on Operational Research Applied to Health Services 1999, Health Statistics and Medical Technology Agency, Riga 1999.

Riley, J. (Ed), Planning for the Future: Health Service Quality and Emergency Accessibility, Proceedings of the 26th Annual Meeting of the EURO Working Group on Operational Research Applied to Health Services 2000, Caledonian University Press, Glasgow, 2001.

Rauner, M., Heidenberg, K. (Eds), Quantitative Approaches in Health Care Management, Proceedings of the 27th Annual Meeting of the EURO Working Group on Operational Research Applied to Health Services 2001, Peter Lang, Frankfurt am Main, 2003.

De Oliveira, M.J.F. (Ed), Accessibility and Quality of Health Services, Proceedings of the 28th Annual Meeting of the EURO Working Group on Operational Research Applied to Health Services 2002, Peter Lang, Frankfurt am Main, 2003.

Dlouhy, M. (Ed), Modelling Efficiency and Quality in Health Care, Proceedings of the 29th Annual Meeting of the EURO Working Group on Operational Research Applied to Health Services 2003, Prague, 2004.

Lagergren, M. (Ed), Quantitative Methods in Aged Care Planning and Operations, Proceedings of the 30th Annual Meeting of the EURO Working Group on Operational Research Applied to Health Services 2004, Stockholm, Sweden, 2006.